Diseases of Poultry

Diseases of Poultry

by

D. Thyagarajan, Ph.D.

Director
Centre for Animal Production Studies
Tamil Nadu Veterinary and Animal Sciences University
Madhavaram Milk Colony
Chennai - 600 051

SSPH

SATISH SERIAL PUBLISHING HOUSE
403, Express Tower, Commercial Complex, Azadpur, Delhi-110033 (India)
Phone : 011-27672852, Fax : 91-11-27672046
E-mail :info@satishserial.com, hkjain1975@yahoo.com
Website : www.satishserial.com

Published by :

SATISH SERIAL PUBLISHING HOUSE
403, Express Tower, Commercial Complex, Azadpur, Delhi-110033 (INDIA)
Phone : 011-27672852 Fax : 91-11-27672046
E-mail : info@satishserial.com, hkjain1975@yahoo.com

Print Again : 2023

ISBN : 81-89304-93-3

Composed, Designed & Printed in India

Preface

Diseases in poultry are the critical factors in deciding the economic viability of the industry. Health of the birds depends on concurrent approach in efficient management, elite germplasm and optimum nutrient input. In the farm, any mortality should be viewed seriously. Dead birds should be subjected to systematic autopsy examination. Health of the flock could be maintained very well, provided the technical input involves scientific approaches in everything. Currently, the poultry industry is not facing any serious set back due to diseases.

Poultry disease diagnosis is an art, involving critical investigation approaches, application of scientific knowledge and further confirmation of the disease. Being specialized in the poultry diseases, I had vast experience in the health management. I am very much confident to diagnose any disease or to face and control any out break. In the diagnosis of the diseases in poultry, autopsy of the birds should be conducted in all the system. All the macroscopic lesions should be pooled, supported by laboratory findings to arrive at exact diagnosis. Further it is imperative, for any disease outbreak; if the diagnosis of disease and treatment are appropriate then it is possible to scale down the effects within 24 hours.

My ambition is just to document all my practical experience into a form of text book for future reference by anybody. This book deals with all about the diseases, predisposing factor, health management, creating immune flocks and basic information. Information furnishings are pertaining to all diseases commonly seen in our country. Accordingly control measures are proposed to suit our condition. Most important pathogonomonic lesions commonly seen here are reproduced. Still continuous updating of the technical information is necessary to make perfect diagnosis.

I hope students, industry people and research scholars will be immensely benefited by this simple document.

Chennai
28-7-2010

Author

Contents

Chapter-1

Introduction

Infectious disease is the result of the invasion of a host by a pathogenic organism. The continued survival of infectious agents with or without the induction of disease depends on their successful transmission to a susceptible host, the instigation of an infection therein and replication of the agent to maintain the cycle of infection.

The three critical factors which favour the transmission of infection include:

- Characteristics of hosts
- Characteristics of pathogens
- Effective contact

A host's susceptibility and infectiousness determine its ability to transmit the infection. Susceptibility to infection should be limited to a single species or group of species. Susceptibility within a species should vary markedly and will be associated with selection of genetically resistant animals following exposure to an infectious agent.

Infectiousness refers to the duration of the period when a bird is infective or the relative amount of an infectious agent that a bird can transmit. A bird is not infectious as soon as it is infected. A period of time elapses between infection and shedding of the agent. This is a parasite's prepatent period, a virus's eclipse period and a bacterium's latent period. In contrast incubation period is the period of time between infection and development of clinical signs. Thus inapparent infections have a prepatent period, but do not have an incubation period. The generation time is the period between infection and maximum infectiousness. These periods, for a given agent and host species are not the same for all animals but show natural variations. Diseases with short incubation periods run a clinical course, terminating in either recovery or death, relatively quickly. Thus a high density is required to ensure that the agents life cycle can be perpetuated.

Three important characteristics of pathogens that affect transmission of infectious agents are infectivity, virulence and stability. Infectivity relates to the amount of an organism that is required to initiate infection. When an agent is capable of infecting more than one species, its infectivity for different hosts is often quite different. Virulence also affects transmission and can change. Repeated passage through the same species of animal tends to increase virulence for that species but to simultaneously lower virulence for the original natural host.

Effective contact describes the conditions under which infection is likely to occur. For a particular infection, it depends on the stability of the organism and the routes by which organism leaves an infected host and enters a susceptible one. Effective contact may be very short or potentially of many years duration. The duration of infectiousness determines the number of susceptibles that can be infected by an infected animal. Behaviour, which may be changed during infection, can also affect the likelihood of effective contact.

The site or sites by which an infectious agent gains entry to a host and by which it leaves the host, are the agent's route of infection. Infection via mouth is one of the most common routes of entry, especially in relation to enteric organisms which often escape from an infected animal in the faeces. Agents that gain entry to the body orally may be disseminated from the infected animal by a variety of routes. The respiratory route is also a common method of transmission for many infectious agents, including those that are not restricted to the respiratory tract. Other routes of infection also include ocular route and skin.

Various factors are responsible for faster commercialization of poultry:

- It is highly prolific in nature
- Shorter generation interval
- It is adaptable to various agro climatic conditions
- Chicken can be reared under high stocking density
- Land needed for poultry need not be fertile
- Water requirement is also less when compared to other species of livestock.
- Poultry can be managed under varied systems, and also it easily adapts to automation.
- Among the various species of livestock, the dressing yield of poultry is the highest.
- Once the skin is removed, poultry meat is the best lean meat suitable for even the sick and convalescing patients.
- It provides more than 20% protein and less than 3% fat.
- The egg has a well balanced amino acid profile with required minerals and vitamins.
- Poultry meat and egg are the food materials which are universally being accepted without being forbidden by any religious taboos.
- Poultry meat and egg are the cheapest food sources of animal protein
- Unlike other animal fats, egg and chicken lipids are not fats, but are oils, good for health. They contain more omega 9 fatty acid MUFA which increases good HDL cholesterol. They also contain considerable amounts of omega 3 fatty acids which reduce the serum bad LDL cholesterol.

Due to above facts, the Poultry industry is developing as mega industry with much diversified activities. Owing to positive development, there is necessity to establish strict bio-security measures to prevent and control the diseases. As such, this book deals in details various practical aspects of the poultry diseases.

❑❑❑

Chapter-2

Avian Physiology

Introduction

The Zoological classification of birds

Kingdom — Animalia

Phylum — Chordata

Subphylum — Vertebrata

Class — Aves

Subclass — Neornithes

Order — Galliformes, Anseriformes, Columbiforme, Parasaiformes

Family — Meleagridae, Phasianidae (Galliformes); Antidae (Anseriformes)

The genus and species classification of domesticated birds are

Chickens	=	*Gallus gallus domesticus*
Japanese quail	=	*Coturnix coturnix japonica*
Duck	=	*Anas platyrynchos*
Goose	=	*Anser anser*
Guinea fowl	=	*Numida meleagris*
Pigeon	=	*Columba livia*
Turkey	=	*Meleagris gallopavo.*

Morphology

The main parts of the body include head, neck, trunk and tail. Head consists of a comb, horny beak, nostrils, and a pair of eyes and ears.

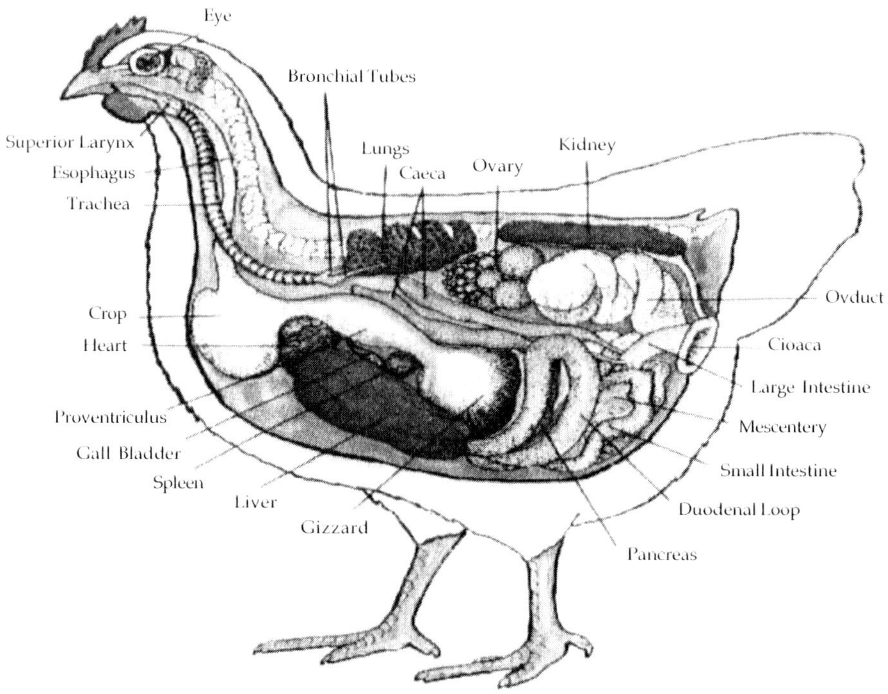

Skin

The secretory glands in the skin are present in the outer ear canal and cloaca. The glands present in the cloaca are called as uropygial or oil gland (preen gland). It is located on the base of the tail which acts as a water repellent mainly in the aquatic species.

Feathers

The root is called calamus. The base is known as quill which gives rise to a structure known as shaft. The shaft again terminates in a structure called rachis. In male, tail feathers are sickle shaped. In drake there is a curved tail feather (sex curl) known as the 'Drake feather'. It is helpful in identifying the sex of the bird. Feathered areas are called as Pterylae whereas areas devoid of feathers are called as apteria.

Five type of feathers are observed in the fowls viz

- Contour feathers – forms the general covering of the body and wings.
- Plumules
- Filoplumules–hair like structures
- Down (Chick)
- Bristles

Comb Pattern

Comb patterns are important taxonomical structure helpful in identifying various breeds and varieties of chicken. The comb is an indicator of the reproductive ability of the bird, as its growth is controlled by oestrogen and androgen hormone levels of the bird. There are many types of comb pattern

- Single comb.
- Rose comb
- Pea comb
- Cushion comb
- Butter cup comb.
- Strawberry comb
- V- Shaped comb.

The external body integrity of the birds is maintained by the two physiological systems:

- Skeletal System
- Muscular System

Skeletal System

Poultry are bipeds. The other pair of limbs of the tetrapod structure are converted as the wings. The skeleton system acts as a basic hard structure to protect the vital organs. Few of the long bones of birds are classified as pneumatic, that is, interconnected with airsacs and bronchial tubes. These air spaces make the bones lighter to help in flight. Certain bones like humerus also assists in respiration.

Muscular System

The muscular system of birds consist of

- Smooth muscle - involuntary in action
- Cardiac muscle – involuntary in action
- Skeletal muscle – voluntary in action

The skeletal muscle consists of three different muscle fibers like

- Red fibers – red fiber is a major constituent having large quantities of myoglobin and fat.
- White fibers - white fibers predominate having higher content of glycogen.
- Intermediate contains both red and white fibers.

Circulatory System

Heart is having four chambers.

I. Right atrium
II. Right ventricle
III. Left atrium
IV. Left ventricle

Erythrocytes in birds are nucleated with mitochondria and endoplasmic reticulum features. Generally, life span of chicken erythrocytes is from 28 to 35 days. Males have higher percentage of erythrocytes than females.

Blood Volume

Young chicks (1 Week old) - 8.7% of body weight

At maturity - 4.6% of body weight

Heart rate in Birds

Type of Bird	Heart rate (Beats/Min)
Canary	1,000
Quail	500-600
Chicken	350-475
Turkey	200-275
Pigeon	220
Goose	200

Respiratory System

The respiratory system starts from external nostrils connected to the internal nares and then to the glottis. The glottis leads to the larynx which is connected to a sound producing organ called syrinx. The long trachea consists of complete rings which divides into left and right bronchi. The bronchioles further extend to form nine air sacs which are as follows:

- Paired cervical air sac.
- Unpaired median clavicular air sac.
- Paired thoracic air sac.
- Paired abdominal air sac.

The lungs do not expand or contract upon inspiration and expiration because of the absence of diaphragm.

Digestive System

The food stays in the mouth for a short time. Upper horny mandible is attached to the skull. Lower horny mandible is divided by a slim slit in the centre that opens to the nasal passages .Soft palate is absent. Tongue is carving knife shaped. Rough surface at the back force food into the oesophagus. Around 24 taste buds are present but however sense of taste is poor. Saliva plays a vital role as a lubricant, for easy food passage. Food passes from oesophagus to crop and then to the proventriculus. Crop accumulates the food for shorter period. There is no enzymatic action in the crop. In crop, feed particles are mixed with mucus. Proventriculus (true stomach) connects oesophagus and gizzard. In proventriculus, gastric juice is produced which contains hydrochloric acid and pepsinogen. Food stays here for a short time.

Gizzard is a round, hard muscular structure present in between proventriculus and upper S.I (duodenum). It contains two pairs of powerful muscles. There is no enzymatic action. Digestion continues by the secretions of proventriculus. Main function of this organ is grinding by using the hard materials like grit, rock present inside the gizzard. Small intestine is having three distinct parts as duodenum, jejunum and ileum which are not very distinct. Various enzymes are secreted in pancreas and intestinal walls. Both digestion and assimilation are important activities. The epithelial lining of S.I has villi for absorption of nutrients. Caeca is represented by two blind pouches. Specific function of caeca is the digestion of fiber and water resorption. In birds, colo-rectum is the combined structure. Cloaca is the common outlet for excreting the digestive, urinary and reproductive products. The external opening of the cloaca is called vent.

Supplementary organs of digestive system are pancreas and liver. Pancreas is located at posterior end of duodenum by way of pancreatic ducts. Liver is represented by two lobes. The right bile duct is extended to form the gall bladder. Bile contains taurocholic and glycocholic acids which are helpful in digestion and absorption of fats and fat soluble vitamins.

Excretory System

Kidneys are bigger in size and are located in the back bones. It is the main control centre for maintaining the proper osmotic balance of body fluids. In birds, the main end product of nitrogen metabolism is uric acid.

Reproductive System

Male reproductive system

The male reproductive organs are two testes present in dorsal regions of upper kidneys. The testes are connected by epididymis and vas deferens to the rudimentary copulatory organ. The structure of the testes is ovoid in nature. The main fuction of the seminiferous tubules is spermatogenesis where as maturation takes place in epididymis. Seminal vesicles and prostrate glands are absent in birds. The sperm head is long, sickle shape with pointed acrosome. It has short midpiece and a long tail.

Semen Yield and Sperm Concentration

Chicken: 0.2-0.5 ml /ejaculate (whitish colour)

Turkey: 0.5 ml /ejaculate (yellowish)

Chicken and Turkey = 1.75 – 3 billon sperms/ejaculate.

Female reproductive tract

In hens, only left ovary is functional, it is situated in the body cavity attached to the backbone. Usually 3,600 to 4,000 tiny ova are present in day old female chicks. This ova matures and develops into full sized yolk.

Infundibulum : After release of ova from the follicle, it is being picked up by the infundibulum. In the presence of live sperms fertilization occurs.

Magnum (Albumen Secreting region) : Thick white secretion is secreted and deposited around the yolk. In this differentiation of chalaziferous layer, the chalaza, and inner thin and thick white takes place.

Isthmus : Inner and outer shell membranes are formed. This membrane primarily protects the egg and its contents from outside contamination.

Uterus (shell gland) : Water and minerals are added to the white through the shell membranes. Here, calcium is deposited over the shell membranes which are formed as egg shell. Colour pigment oocyanin/ oporphyrin is added for the egg shell colour.

Vagina : Before laying, the egg passes to this section.

❑❑❑

Chapter-3

Immune System of Domestic Fowl

Immune system protects the body against microbiological and environmental disturbances. It protects the body from pathogens like bacteria, virus and parasites. It is generally referred as 'Ministry of defense'. Antibodies are the best defense mechanism against infections. Bird requires a competent immune mechanism to counteract the effects of infections and vaccinations.

Immunity

Immunity is coined from the Latin word 'Immunis' which means free or exempt.

Types of Immunity

It has two components.

- Innate, and
- Acquired

Innate Immunity

It is seen before birth itself. This word comes from Latin word 'innatus' which means to be born with. It is a slow and non-specific immunity. It is having three components namely anatomic barriers, physiological barriers and cellular components.

Acquired Immune System

Acquired Immune System is specific for its specificity, hetrogenecity and memory. It is classified into non - cellular and cellular component.

Avian Immune Response

Immune response to infectious agents is by production of antibodies and by lymphocytes. This is referred as humoral immunity, likewise the immunity induced by cells are classified as 'cell mediated immunity'. These two are the important segments of acquired immune system.

In antibody-mediated response, protections from antigen are provided by engulfing the antigen by the body's phagocytic cells namely neutrophils, and macrophages. Antibodies are two types like extra cellular and intracellular. Cell mediated immunity comes into action when infectious agents are uncared by humoral immunity.

Anti-body Mediated Immune Response

The chicken blood usually have two important type of cells namely red blood cells and white blood cells. Normally lymphocytes constitute 60% of the white blood cells. Lymphocytes are produced in the bone marrow. Bursa and thymus are the two target organs for primary lymphocyte production.

Lympocytes are classified as B.cells and T.cells. B cells are produced from bursa of fabricius and T.cells from thymus. These are the secondary lymphoid organs.

Action of Lymphoid Stem Cells in the Bursa

Primarily lymphoid stem cells are differentiated into matured B lymphocytes in bursa. Bursa usually atropies or regresses after sexual maturity. Afterwards secondary lymphoid organs like spleen, caecal tonsils, payer's patches, meckel's diverticulum, plays a role.

Mechanism of Producing Antibodies by B-cells

B cells response positively to repeated exposure of antigens and infectious agent, and secrete large amount of antibody which are called as 'plasma cells'. Antibodies are known as 'immunoglobulins (Ig), which are proteins made up of amino acids. Each plasma cell is capable of producing 300 molecules of antibodies per second.

Prevention of Auto-Antibody Production by B-cells

Each B-cell is first screened by the bursa. Bursa then selectively destroys such B cells that react against the body's own components. This mechanism is known as 'negative selection' or 'clonal selection'. Upto 90% of the B cells are thus eliminated through a process known as 'apoptosis' or physiological cell death'.

Antibody diversity' is the process of maturation and differentiation of lymphoid stem cells. One specific type of B-cell produces antibodies only against one particular infectious agent.

The process is extremely complex. B-cells carry tiny ready made antibodies on them, against each infectious agent / antigen present in nature. The antibodies (Ig) act as receptors for infectious agents and are known as B-cell receptors (BCRs). When bound to B-cell surface, this antibody molecule acts as an antibody.

Each B-cell is covered with about 2,00,000 to 5,00,000 antigen receptors. The infectious agent picks up those particular B-lymphocytes from the birds body.

Reaction of antibody against the infectious agent

The antibody molecule looks under a high power microscope. i.e. "Y" shape letter. On the arms of 'Y' are the areas which recognize the foreign invader. The situation is like a lock and key, where the portion of the invader (key) matches with a portion on the antibody (locks).

Recognition of the Foreign Invader by the B-cell

In 'Innate immune system', the macrophage can engulf infectious agents and destroy them by powerful enzymes. Cells, like macrophages can ingest, process and present an infectious agent to B-cells is called as 'antigen - processing (presenting) cells'.

Types of Antibodies

In chicken 3 main types of antibodies (Ig) are produced. They are IgM, IgG, and Ig A.

IgM, is the first antibody produced in the chicken which is actually five of the Ig's (IgY) joined at the tails in a circular fashion. The function of IgM is mainly to arrest the entry of infectious agent. It prevents the infectious agent from multiplication and causing damage. This process is called as "agglutination'. IgG is the next antibody synthesised. IgG consists of only one 'y' and hence five times smaller than IgM. IgG is measured by enzyme – linked immuno sorbent assay (ELISA) test. Immunoglobulin cause clumping of the infectious agents and attach to their surfaces. This facilitates phagocytosis. IgG is found in highest concentration in blood. It plays major role in antibody- mediated defense mechanisms. It is the smallest of three antibodies and hence IgG escape from the blood vessels. IgA is the third important antibody in the chicken. It is manufactured by B-cells present in tissues just under the body surfaces. It is also produced in epithelial cells which diffuse into blood stream and into external secretions. IgA is produced mainly to protect body surfaces. This process is known as "Immune elimination' or exclusion'. IgA can therefore act at three different points to restrict infectious agents in the intestinal wall, in the lining intestinal cells, in the intestinal lumen. In chicken 30% to 75% of IgA produced in the intestinal wall diffuses into the blood circulation for reaching the liver. By this mechanism infectious agents are eliminated from the body.

Parental Transmission of Antibodies

A day-old chick has an inadequately developed immune system and therefore depends on the immunity from the mother. Newly hatched chick has temporary immunological protection until systemic production of antibodies begins IgG in yolk is having equal antibody titer as hen's blood. IgM and IgA from oviduct secretions are found in the albumen. IgM and IgA from the albumen diffuse into the amniotic fluid and entered into the embryo.The function of maternal antibodies is to prevent pathogenic organisms from causing disease in young chicks.

Cell Mediated Immune Response

Lymphoid stem cells enters into the thymus where in differentiate into mature lymphocytes. This differentiated mature cells are referred as "thymus – derived lymphocytes' or T.Lymphocytes or T-cells. Further processing and differentiation of this cells leads to 'cytotoxic T-cells. By this, they acquire the ability to kill the cells which harbour the infectious agents.

Purpose of the Cell-Mediated Immune Response

The B-cell antibody system is meant basically to destroy the infectious agents before they reach their targets. T-cells system is to deal with the infectious agents after they reach their target cells.

Cells are several types as

- Cytotoxic T-cell
- Helper T-cell
- Suppressor T-cell.

Action

Viruses within the cells, bacteria and other infectious agents are destroyed by cytotoxic T-cells which protect the body from any furthur damage. Cytotoxic T cells have receptors on their surface as lock and key mechanism. Each cell possess about 10,000-20,000 similar receptors.

Action of Cytotoxic T-cell over the Infected Cell

Within a few minutes, cytotoxic T-cells bind to infected cell, leads to the release of granules. From the granules perforins and enzymes are released. These enzymes activate the 'Suicide enzyme system'. Thus the infectious agents are killed.

Macrophages in Immune Response

The Infectious agents are first engulfed by a cell called 'macrophage'. It is derived from blood cell known as 'monocyte'. Infectious agents are broken down, processed and then presented to T-cells. Macrophages thus act as 'antigen-processing' or 'antigen – presenting cell.

Ways to Protect Chicken Against Infectious Agents and Improve Immune Response

- The immune response is a complex chain of events. The basis of any programme is to ensure that birds are protected in an effective and well implemented biosecurity programme.

- In innate immune system, protected the birds are by best management. The innate immunity is intact and acts as a functional barrier to restrict invasion.
- When birds are forced to breathe poor quality air which are high in ammonia and dust, the defense barrier gets break down.
- The same holds good for feed and feed quality which contains mycotoxin. The local defense mechanism again gets break down and make a bird easily prone to intestinal infections.

❒❒❒

Chapter-4

Genetic Resistance to Poultry Diseases

Recently much interest has been shown to study the interactions between genetic properties of the host and disease agent in relation to the development of the disease. Likewise more importance is also given on genetic aspects of disease resistance in poultry breeding. Consequently, some of the major genes have been identified which play a major role in resistance of poultry especially against tumour diseases of viral origin and control of immune responses.

In general, there are three basic approaches for the disease control which are (1) eradication of disease (2) vaccination or medication and (3) improvement of genetic resistance to disease. The adoption of first method may result in culling of innocent one, because some of the individuals in a flock are showing the disease. Vaccination or medication can be adapted mostly to the infectious diseases, whereas the genetic resistance to diseases give protection against heavy losses during the breakdown of immunity and also during drug resistance. It is also critical to be aware of the status and level of the genetic resistance to the diseases in a flock. Now the involvements of the genetic resistance to the diseases become one of the main parts in the disease control.

The genetic resistance to diseases in poultry can be classified as follows.

1. General disease resistance
   (i) Non specific mean
   (ii) Immunological means
2. Specific disease resistance

General disease resistance to diseases is capable of defending against the invasion by the varieties of pathogen. It should protect the birds against all the diseases caused by all the living organisms and also against stress, which is a main predisposing factor. It need not give protection to diseases or disorders caused by non-living agents such as chemical poison, feed deficiencies, trauma, endocrine disorders, autoimmune disorders and other abnormalities. It is also necessary to prevent the production losses, which may be caused by morbidity and by subclinical infection. Generally, the resistance is manifested by two ways, one is by ability to resist infection and another is by ability to overcome or to live with the disease, among these the later one is the most desirable.

Indicators of General Disease Resistance

The general disease resistance is an ability to resist any alteration in the state of the body by external causes like microorganisms and from stress, which alters performance. Since it is difficult to find out the general disease resistance in birds, as it involves exposure of pathogen, its existence are correlated mostly with other traits.

(i) *Non specific means*: This seems to be potential source for increasing the general disease resistance. It is well known that the hybrid vigour is particularly important for viability of the chicken. The resistance to disease depends not only on the paternal or the maternal line but also on the age, sex, strain, line and breed of the animal and also on the type of the infectious agent as well. The enzyme lysozyme present in most animal tissues and fluids is useful in protection against a wide spectrum of bacteria. Interferon which is species specific, inhibits viral replication, is another potentially important component of general disease resistance. Low levels of plasma corticosterone were not favourable for resistance to bacterial infection but result in increased resistance to mareks disease, new castle disease and mycoplasma gallisepticum. It is also reported that by administering the corticosterone, the resistance to infestation of the parasites is increased.

□□□

Chapter-5

Transmission of Diseases

Diseases are mainly predisposed by stress and disturbance in physiology, which could be due to so many factors. After infection, organisms produce multiple toxins, which affects the health of the birds. Intensity of the diseases varies with virulence of the organism, number, type, mode of entry, taxonomy and immunity of birds. Producer need not bother about diseases, if proper preventive measures are taken. Appropriate preventive measures are possible with thorough knowledge on transmission methods and disease affecting methods.

Infectious diseases are caused by so many types of organisms. Infectious organisms can be classified as bacterial, viral, parasitic and fungal organisms. Entry of disease causing organism from infected birds to uninfected birds requires intermediate materials. Such intermediate materials can be classified as three types.

- Air
- Water
- Miscellaneous

Air

Air borne transmission is possible with smaller sized organisms. This type of organism can reach from one place to another by floating. Moisture present in air assists in transmission of diseases. Mostly virus is transmitted by this way. Air velocity and direction of flow decides the intensity of the diseases outbreak. Direction of airflow is yet another important factor deciding the transmission of diseases.

Drinking Water

Most of the diseases in poultry farm are transmitted through water. Drinking water and waterers act as agencies for transmission of diseases. Moisture is essential for multiplication of disease. Diseases transmitted through drinking water induce outbreaks of higher intensity. Healed tissues, mucous discharge and excreted materials of infected birds usually contaminate the water and transmit the diseases.

Parasitic diseases such as worm infestation and coccidiosis are easily transmissible through water. Such organism egg/oocyst is usually discharged through droppings, which easily contaminate the drinking water thereby transmits the disease.

Miscellaneous

Transmission is also possible with intermediate materials of live or inanimate objects. The major intermediate materials are.

a) Human

Farm managers, farm labours and farm suppliers are responsible for mechanical transmission of diseases. Leg shoes and clothes frequently carries the organism from one place to another. Discharges and excretory products of infected birds usually attached to the floating materials and are transmitted.

b) Neighbourhood Farms

In general, during the outbreak of diseases, the person from infected farm should be restricted to visit the neighbourhood farm. Even during normal period, it is better to avoid entry of outside visitors.

c) Technical Manpower

Technical manpower like blood collecting personnel, debeaker, vaccinator and sexer are frequently attending their work in the farm.

In case, if these personnel enter many farms, there is every possibility that they act as agents for transmission of diseases. Technical manpower should take necessary precautionary measures to control transmission of diseases.

d) Vehicles

Transport vehicles supply requirement of the farm and take delivery of the farm products. Chick and feed supplier requires vehicle for easy mobility. However, the vehicles are major places for transmission of diseases.

e) Recovered Birds

Recovered birds continue to discharge the infectious organism. Histomoniasis, leucocytozoan, coccidiosis, pullorum disease, tuberculosis, fowl cholera, lymphoid leucosis, salmonellosis, fowl pox and many other diseases are spread from affected and recovered birds. It is advisable not to bring the aged birds from one farm to another. Recovered and unproductive birds should be culled at once

f) Egg borne diseases

Some of the diseases can be transmitted through eggs.e.g.Pullorum disease, adenoviral infections etc. It is advisable to have the parent stock free from these diseases.

g) Equipment

Farm requires more usage of equipments. Such equipments are usually parked with litter material, through which it is possible for spreading of diseases. Egg transfers, weighing balance, chick baskets and gunny bags usually carries infectious organism of coccidiosis and fowl pox.

h) Rat

Many diseases are transmitted by rats. Rats by way of its excretory products like urine and dropping contaminates either feed or water. Usually the rats are coming in search of feed. Feed drums and storage places should be properly protected from rats. Existence of salmonellosis in farm is mainly due to rats.

i) Birds

Flying birds like crow and sparrow generally act as sources for many diseases. Water storage and feed storage bins should not be freely exposed to open sky. Faecal dropping of free flying birds usually transmits the virus of ranikhet disease.

j) Fly

Domestic fly is a major problem in the farm. Farm hygiene can be accounted by assessing the number of domestic flies. Cage system of rearing predisposes fly nuisance. Domestic fly eggs carry the infectious organisms. When the birds sit on the feed and water, it gets contaminated and induces disease outbreaks. Parasitic diseases caused by worms and coccidia are easily spread by this way.

k) Feed

Quality feed is essential for the farm. Many feeds have disease causing organisms. Decomposed fish contains pathogenic organism. Procurement of animal protein should be of good quality and should be free from pathogenic organisms. In rainy days, fungus grows in the feed resulting in the production of fungal toxins. In the event of presence of microbial organisms, usage of anti bacterials or antibiotics counteracts the ill effects.

l) Laboratory examination

In the course of disease investigation, affected birds are sent to laboratory for isolation and identification of organism. Such birds harbour lot of infectious organism and also infected with other pathogens present. In any case, these birds should not be brought back to the farm. Diseases are spread by means of so many ways. Farm hygiene and disease preventive measures decide further disease outbreaks. Some of the points that should be considered positively are,

1. Use strong disinfectant at the farm entrance
2. Litter should be properly discarded
3. Stagnation of water should be avoided
4. Avoid usage of decomposed fish
5. Eliminates high moisture content in any place
6. Fly and rat nuisance should be avoided
7. Entry of visitors and other farm labourers should be checked
8. Cleanliness of waterers should be maintained
9. Proper well planned vaccination processes
10. Entry of flying birds should be checked

□□□

Chapter-6

Microbes-Secondary in Poultry Diseases

The development of the poultry industry is very rapid in recent days. The birds are not an egg producing machines but they are giving nutritious and protein rich eggs and meat for human consumption. The farmers are rearing their flocks in the poultry sheds in large scale without giving due attention to the advanced scientific techniques. Hence, the birds are likely to be succumbed to the diseases. The causative factors of the diseases are many. But the farmers are thinking that the birds are affected mainly and mostly by the microorganisms. Actually the microbes take upper hand and cause the disease, only if there is deviation in the primary factors such as management, environment, nutrition, hygiene and disease control. Some of the main primary factors for the various disease conditions in poultry are,

1. Problems in Management.
2. Problems in Environment.
3. Problems in Nutrition.
4. Problems in Hygiene.
5. Problems in disease control.

The above factors reduce the vitality of the birds and set the dormant microbes to take upper hand and cause the disease in birds. Whenever if there is any abnormality in the birds, the farmers should contact immediately the veterinarians and make correct diagnosis and give proper medication. The farmers should ignore the assumption that the birds are suffering mainly by the micro organism and try to avoid incorporation of antibiotics and other chemicals in the feed and water. This creates stress and reduces the vitality of the birds. The farmers should eliminate these misconceptions from their mind and adopt effective scientific managemental practices, giving balanced and nutritive diet, fresh and clean drinking water, proper vaccination and hygienic condition in and around the poultry farm as detailed below; all of which can increase. The vitality of the birds for better production and profit needs effective managemental practices.

The location of a farm has to be selected carefully. The environment around the poultry farm should be clean and hygienic. For rearing the different age group of birds, adequate floor space must be provided. In the sheds, the feeder and waterer numbers and the space requirements should be within the norms. No birds should be allowed to starve due to inadequate feeder and waterer space and number. Since the curtains around the poultry houses hinder the free flow of air inside the poultry houses, the curtains can be removed to reduce the production of ammonia and its accumulation in the houses. Daily checking may be carried out in and around the waterers to remove. The caked litter materials should be avoided so as to avoid the outbreak of coccidiosis in the birds. To overcome such outbreak the litter must be dry always. Most of the farmers specifically in Namakkal area do not aware of the fact that, the cement flooring reduce the disease burden and carry disinfection. The farmers think that the cement flooring increases the cost of construction of poultry houses and satisfied with the mud flooring for least profit with increased expenditure. During the house construction if they invest extra small amount for cement flooring they can enjoy more profit with less expenditure for medicines and more satisfaction because of their healthy birds. The cement flooring destroy the micro organisms, vectors and parasites effectively and can grow healthy replacement stock. In mud floor, despite the difficulties in cleaning operations the parasitic eggs and vectors hiding inside the mud floor can escape from the disinfection operations while constructing the poultry houses. The length of the poultry houses can be any length but the breadth should be within thirty (30) feet for better ventilation, effective watering and feeding management.

In the chicken body about 70% is water. The nutrients for the metabolic activities of various organs and the waste products to the excretory organ are transported through the water medium. Hence, the most important role of the poultry farmer is to provide clean and fresh water to the birds for better health. The unclean water is enough to deteriorate the health of the birds. In drinking water, many of the useful minerals and some of the harmful minerals are also present. The water must be analysed before use to avoid the presence of harmful minerals in drinking water. For example, if the sodium chloride is present in excess in water it kills the younger birds or produce watery dropping in older birds. Moreover, in laying birds the thickness of the eggshell likely to be lowered and naturally the breakage percentage may increases in the farm. In unclean water the microorganisms multiply quickly and these organisms affect the birds so the vitality of the birds goes down. When the farmers use the disinfectant like chlorine and iodophores in drinking water it is possible to control and destroy the microorganisms present in the drinking water. This type of disinfectant does not give any problem to the birds. Two third of the waterers may be filled to avoid any spillage of water in the litter, during consumption. In summer, the water can store in mud pots and ice cubes may be put in the waterers to avoid summer stress. The waterers must be filled with clean water throughout the day.

Care must be taken to give good quality feed to the birds. Balanced and nutritive diet has to be prepared for the respective period of growth. Feed samples need to be collected from the every consignment of feed bags and submitted to chemical analysis for nutrient requirements. If there is any nutrient deficiency immediately incorporate those particular nutrients for better production and vitality. Good quality vitamins and minerals should be purchased. If the quality of the feed ingredients, feed additives and feed supplements are not good the birds suffer from deficiency. Diseases inturn the vitality goes down. The feeders must be filled frequently throughout the day. The rations have to be computed according to the seasonal requirements. The feed must be analysed for excess sodium chloride. These elements cause the mortality and lower the vitality of the birds. Further feed samples must be seen for the presence of fungal growth. The fungus grows rapidly in the wet rations, oil cakes and maize. These fungi produce mycotoxins and these toxins induce mortality. If the quality of the feed is lowered or if the nutrients are inadequate in the feed, it leads to lowering of the vitality of the birds and the dormant microorganisms take upper hand and cause the disease. For better profit, healthy birds and the quality of the feed is very important.

Effective Vaccination Programme

The bird has to be vaccinated against the viral diseases because there is no treatment for viral diseases. For effective immunity development, the vaccines must be handled carefully. The farmers are least bothered about the preservation of the vaccines. This is not a good trend. The vaccines should be preserved in ice cubes or deep freezer until it is used. The vaccines should be transported to the farm in a thermoflask with ice cubes. But nobody is bothering about the preservation of the vaccines. Naturally the required level of immunity may not set in. The birds should be vaccinated only by the qualified veterinarians. Moreover, the vaccination should be done at the appropriate age as otherwise the immunity break down may occur. Further, if the flock encounter with mortality, the death birds must be subjected to autopsy examination, the diagnosis must be made; proper medicines should be given with help of veterinarians. The medicines should be given with proper dose and the period of treatment must be followed. Unnecessary medication and inadequate dosage should be avoided.

From the above points, the microorganisms are secondary in causing the diseases. If the managemental practices, feeding and watering management and vaccination programmes are effective the microorganisms cannot set up the disease.

❑❑❑

Chapter-7

Poultry Disease Investigation

Poultry disease investigation is an art of methodical approach for finding out the cause. Nowadays, the etiological diagnosis of the poultry problems in the field is not so easy, as that of earlier days. Investigators approach must be multifaceted approach of environment, nutritional, managemental, genetic causes. Major problem can be attributed to specific causes easily. However, minor problems like 5% drop in production, shell quality faults, fall in hatchability, leg weakness or loss of performance are difficult to find out the cause. These are the minor problems, wherein methodical investigation is necessary to pinpoint the problems. Based on these factors, investigation procedure can be classified into two major headings.

1. Etiological diagnosis based on the postmortem report, subsequently confirmed in laboratory.

2. In depth investigation approach to find out exact cause.

It is mandatory to develop specific plan of investigation procedure, which will help to identify the cause even in different and difficult situation.

Identifying the Cause

In general, problems are due to deviation from quantifiable parameters. It could be mortality to variation in the shell colour. In that case, it is necessary to have the normal value of various categories of the poultry. Variation in production between different classes of poultry and management is normally expected. Normal standard should be referred, to arrive conclusion on deviation from production or otherwise. Technical staff expertise is of great importance in arriving the exact cause for the deviation in health or production.

History Taking

This is one of the important steps in identifying the problem. Process of history collecting is the way of creating check list of various factors. The approach should be methodical to get answers for the entire question. Some of the answers can be from the farm itself during the course of investigations. Farmer and farm persons should systematically querried for collecting the right information.

Investigation

History collection itself can give a lead to establish exact cause of the problem. Sometimes investigator may have a mindset on the prime cause of the problem, probably associating with age or other factors. In general, autopsy examination of the birds can definitely give an idea on exact cause of the problem. Laboratory examination of the suspected material confirms the etiological factor. Collection of suspected material is a critical factor to arrive a positive conclusion. In the event of autopsy examination, it should be conducted on fresh dead birds. Selective ailing birds should be the representative of the flock having the problem. Usually it is advisable to submit 8-12 birds for autopsy examination to pinpoint the cause.

Necropsy Examination

Necropsy examination of the birds is the best method of diagnosing the disease. It should be methodical so that no other system is omitted from examination. It is advisable to see autopsy examination as many as possible to come out to a correct diagnosis. In depth investigation of the birds initially, followed by confirmatory checks on the subsequent birds are correct steps for positive conclusion.

It is also important to note down the symptoms of the active ailing birds. Before proceeding for autopsy examination, it is necessary to have a look at the bird's external appearance like presence of wounds, colour (icteric or pale) and external parasites. Birds can be dipped in diluted disinfectant solution except head region. Then the carcass should be laid down on the post mortem table.

First angle of the jaw is cut open proceeding downwards through pharynx, esophagus, and crop for internal examination. The trachea from larynx needs to be cut open for observing any abnormalities inside. Before proceeding further, size of the thymus gland need to be noted.

Then leg on either side of the body needs to be separated from body by making incision in the skin. Carefully femoral joint is dislocated removing the femur head from the acetabulum. The joints of the legs have to be examined for any abnormalities. The tendon and synovial fluid are to be seen for any abnormalities. The muscles in the thigh region should be separated to see any abnormalities in sciatic nerve as enlargement or discoloration.

Next, skin in the region of abdomen is cut horizontally and flapped upwards to expose the abdominal region and breast muscle. Breast muscle is the place for seeing signs of dehydration as sticky feeling and congestion. Evidence of gout could also be seen in breast muscle. Surface hemorrhage may be evident in infectious bursal disease. Then, abdominal wall is cut across each side upto clavicle bone. Cut parts are turned upwards and forwards so as to expose the internal organ for full view.

Initially, the internal organs needs to be examined as such the liver should be critically checked for any lesions such as discoloration, enlargement or hemorrhages. The serosal surface of the airsacs needs to be seen for the presence of purulent material. The appearance and changes of the heart and pericardial sac are essential. At this stage using the hot knife/swabs can be taken from the exposed organs for cultural examination without contamination. Later, intestine at both ends have to be cut for examination of mucosa and serosa. The proventriculus should be cut open to see any petechae or haemorrhages in the gastric papilla. The reproductive tract should be seen including genital organs for size, number and conditions. Further other organs like kidney, sciatic plexus and bursa of fabricus needs to be checked for inflammation or changes in size. The myocardium should be cut open to see inside and valves.

The lungs are important organs to assess pneumonia or consolidation. The nasal cavities and sinuses have to be observed for the presence of exudates by cutting the upper beak at appropriate level using sharp scissors.

The brain's external appearances have to be examined. If necessary, materials can be collected for histopathological examination. Incision of all along outer region of skull is necessary to remove the skull and to expose the entire brain for examination.

During the course of the methodical postmortem examination, tentative diagnosis can be easily drawn. However it must be supported with laboratory examination, for which it is necessary that suspected materials are need to be collected for microbiological examination.

Further laboratory examination such as cultural examination, virological tests and parasitological examination, histopathological examination and serological examination are to be conducted for confirmation.

Mortality Investigation

Approach for investigation of morbidity or mortality shall be methodical to find out the cause. If there is mortality during the first week, suspects for yolk sac infection. In the second week, the cause could be chicken anemia agent. Subsequent weeks, the problem could be due to IBD. Elder birds succumb to secondary bacterial disease such as Escherichia coli infection. Environmental and various infections complicates with secondary bacterial diseases. Exact cause confirmation should be with laboratory findings like combination of pathology and other appropriate tests. Recently PCR tests are available for getting reliable and sensitive results. Isolation of etiological agent particularly virus and bacteria are important for confirmation. Antibody based tests are difficult to draw a conclusion due to existence of vaccinal titre or maternal antibody titre, owing to its shorter life. However in event of major outbreaks, some birds should be retained till laboratory results are arrived. It will also give chance to assess of vaccinal or infection titre. Apart from this, SPF birds can be put into diseased birds for picking up the disease for further investigation.

It is equally important to look into other factors like environmental deficiency. During personal inspection of the farm, behavior of the birds are very critical. Huddling together under heat source is easily reflect on the uncomfortable environment. Deviation from feeding habits or making noise can give an idea about the health of the flock.

Investigation of Production Drop

Investigation of finding exact cause of slight drop in production may be very difficult. It could be one cause or multiple causes. However it is advisable to effect as many changes as possible to get back to past production. Even in the course of investigation it is possible to isolate the organism of importance or evidence of titre indicating possible cause. Mostly the problem of drop in production is feed related. The feed related problem should be investigated particularly and firmly. In the case of drop in production, managemental factors like feeder space, waterer space and quantum of light given, water quality and environmental factors are to be analyzed thoroughly.

Problem in Hatchery

It is quite difficult to investigate and find the cause for lowering of hatchery efficiency. It requires investigation of both biological system and mechanical operation. Logical discussion of the problem may corner the cause. Experienced technician can able to pin point the cause. Generally in a flock, the hatchability is low reaches peak and again decline subsequently.

Investigation of the hatchery problem involves thorough checking of the records. It must be seen from the following, the problem is localized or general in nature.

- Breed
- Age of males and Females
- Existing disease outbreak status
- Medication programme
- Egg production trend
- Feed quality
- Management problem like wet litter
- Frequency of egg collection
- Egg sanitation procedures
- Age of egg and storage details

In course of investigation, it is necessary to examine hatchery waste, hatchery environment, breeder stock nutrition and hatchery hygiene to come for right conclusion.

Hatchery Waste

Materials of the hatchery need careful examination. In that unhatched eggs form the main material, the causes of failure of hatching needs to be obtained by break open studies of the unhatched eggs. The embryonic mortality can be listed as follows.

1. Early embryonic mortality
2. Midterm embryonic mortality
3. Late embryonic mortality
4. Infertile eggs
5. Malpositions and Malformation
6. Exploder

This is necessary to examine at least 100 unhatched eggs. In general embryonic mortality reaches peak at 3-5 days (25%) and again at about 18-21 days (50%). It is expressed normal embryonic mortality as 25% during 0-4 days, 10% during 4-16 days and lastly 65% between 16-21 days. In the turkey, midterm embryonic mortality reaches during midterm period of 18-24 days.

Early embryonic mortality is mostly due to maladjustment at the critical period of formation of basic circulatory system. This development occurs between 3-5 days. Usually 30% of early embryonic mortality is due to physical trauma associated with transportation, resulting in loss of air cell, improper egg storage conditions, overheating or chilling, faulty incubation, improper fumigation, poor hygiene and inadequate turning. Biotin and Vitamin E are important nutrients required for the vascular development.

Midterm mortality is mostly attributed to nutritional problems like vitamin B_2 deficiency which induces clubbed down and edema. Micronutrients like phosphorus, zinc and specific fatty acids like linoleic acid are responsible for enhancement of hatching performance during this period.

Late embryonic mortality which accounts for 60% is mostly due to micro environmental faults like temperature, humidity, ventilation and inadequate turning, late transfer, prolonged egg storage and Trans-ovarian diseases. Nutrients like B_2, biotin, folic acid, B_{12} and manganese plays a vital role during this period.

Culls refer poor quality chicks or poults hatched, which accounts to 1%. This is due to malformation (old eggs), weak chicks, machine faults, late hatch) or poor chicks (nutrient). In turkeys, higher incidence is observed than chicken due to unhealed navel which could be created by old eggs, higher setter temperature and humidity. In addition leg abnormalities, underweight, deformity and distended abdomen could also be seen in weak chicks.

Certain drugs like demetridazole, furazolidone, organic mercurial, inorganic mercurial (end products of insecticides), produces deleterious effects. Adverse effects of nicarbazine at low levels are already established.

Disease investigation is an art of finding out the etiological agent. This should be methodical and technical in most efficient way to prevent damage to the birds and drop in production.

❑❑❑

Chapter-8

New Castle Disease

In Western Countries this disease is reported in a place called Newcastle. In these countries, it is known as New Castle Disease. This disease is first recorded in the place of Ranikhet in northern part of India. Since then it is also known as Ranikhet disease. It is a highly contagious disease. In the year 1926 at Indonesia, the disease is seen in New Castle Tyne.

Cause

Virus is of RNA type. It comes under Paramyxoviridae family. Many types of avian paramyxovirus are available. Avian paramyxovirus 1 causes Ranikhet disease which produces much affection. Depending upon the acuteness of the infection, the disease is classified as five types.

The Virus can be easily destroyed by heat, X-ray and many chemicals. Virulence and disease producing capacity may vary with the disease type. The disease manifests in five different forms as.

1. *Viscerotropic velogenic* : Per acute form affects all the internal organs. Much intensive changes are noticed in intestines.
2. *Neurotropic velogenic* : Per acute form affects the nervous system. Symptoms are seen in respiratory and nervous systems.
3. *Mesogenic type*: Moderate variety with less mortality wherein symptoms of respiratory and nervous symptoms are seen.
4. *Lentogenic type*: fewer symptoms. Respiratory system is not affected much.
5. Symptomless new castle disease affecting intestine.

Ranikhet disease attacks 200 types of birds. This disease is contagious in nature. Goose and ducks are affected by highly virulent virus, as such there is not much symptoms.

Transmission

Incubation period of the disease varies from 2 days to 2 weeks. Transmission of disease depends upon the management condition and association of the birds. Organisms enter the healthy birds by feed contamination or by airborne transmission. The virus easily and rapidly spread by air. Disease is spread at a lower rate by the organism through the droppings.

Labourers and farm personnel's also act as transmitting agent of the disease from one farm to other farm. Man directly or by indirectly through equipment, gunny bags, live birds, dead birds transmits the disease. Disease transmission also depends upon the infectivity of the birds. In dead carcasses, viruses are alive for many weeks. Under wet storage, the virus survives for many years. In the faecal dropping, the survivability limits to month in room temperature.

Symptoms

Symptoms are specific to the type of infectious organisms, virulence, and nature of affection and immune level of the birds. Sometimes the disease occurs as mixed infection and causes much damage. In per acute type, the infected birds die without any symptoms. In the intestine type, depression, respiratory system affection, dullness, debility and finally mortality are seen. Dropping is greenish tinged. Eyes and noses are edematous due to involvement of respiratory system. Before mortality, muscle tremors, nervous system disturbances like torticolis, paralysis of wings and legs are observed. Mortality rate reaches upto 100%.

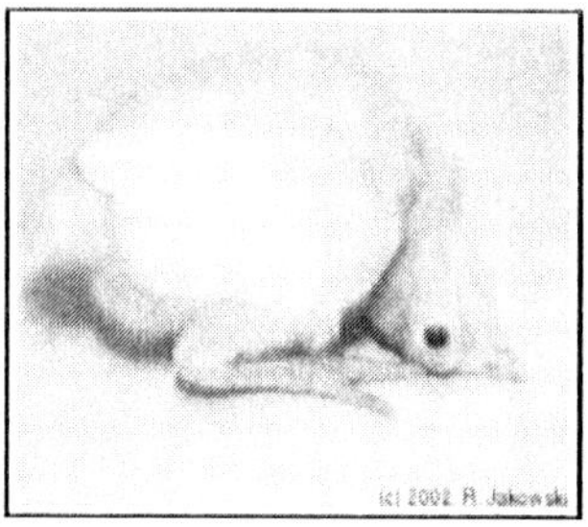

In nervous type, the symptoms of respiratory tract and drop in egg production are seen. Due to effect on egg shell, Shell less egg are noticed. Diarrhoea is not common. In chicken, morbidity rate is 100% and mortality rate is 50%.

Mesogenic type of Ranikhet disease manifests with much respiratory symptoms. Drop in production and nervous system affection are common. Mortality rate is less.

Lentogenic type does not cause disease in grown up birds. In young birds, it causes respiratory system disturbances.

Lesions

Lesions vary with type of disease attack. In the intestinal type, diphtheritic ulcer in the mucous membrane of the intestine is seen. In all type of diseases, petechiae in the papilla of the proventriculus and caecal tonsils are common pathognomonic lesions. In the respiratory type, the hemorrhages are seen in the trachea. A septicaemic lesion like petechiae on the epicardial fat of the heart is present. Immune level of birds is critical factor in inducing the lesions.

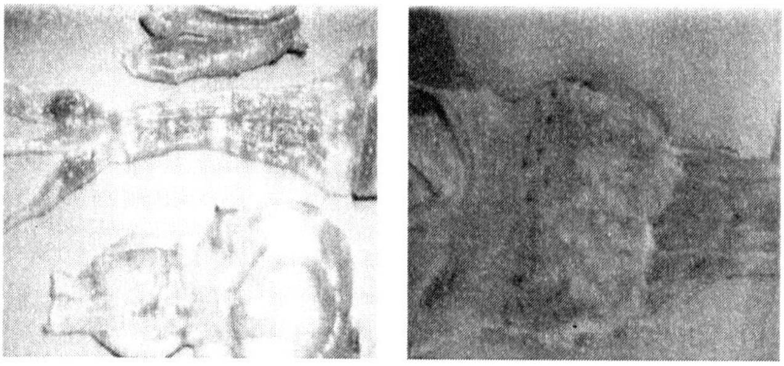

Acute form: Hemorrhage in the mucosa of the trachea (upper), large intestine, particularly caecal tonsils (middle), proventriculus (bottom) and gizzard.

Immunity

Immunity is established by two ways. Both Cellular and humoral immunity are established in Ranikhet disease. Haemagglutination inhibition titre is basic in assessing the immune level of the birds. HI titre of 1034 is required to withstand the infection. In the ranges of 64 to 1034, affects egg production if infection is set in without any mortality. 16 to 64 HI titre induces 10% mortality. Less than 4HI make the birds highly susceptible to infection and mortality rate is 100%.

In the chicken, cellular immunity is the primary factor in preventing the infection. This immunity is established within 2 to 3 days. In vaccinated chicken the immunity is through cells.

Humoral immunity is also responsible for prevention of disease. After infection and three to four weeks after vaccination the humoral immunity is produced. It remains in the systems for a period of one year. Immune level varies with virulence and type and frequency of vaccine.

In the affected places, flock immunity is also created. Local immunity is seen by way of sections of intestines and lungs. This type of immunity differs with incidence of disease. Parental immunity determines the maternal antibody level in the chicks. Chicks are protected from the disease by maternal immunity. Immune level in the birds differs with intensity of the disease. Immuno depression happens by Infections of bursal disease. In the out break of bursal disease, the vaccination does not elicit the immunity. Chicken anemia also produces same type of effect.

Diagnosis

Symptoms, lesion and isolation of the causative factor are the basic criteria for diagnosis. Haemagglutination property of the virus is exclusively used for laboratory investigation.

Prevention and control

Multiple approaches in control of the diseases are necessary. Hygienic environment is the critical requirement. In any possible way, spread of infection from infected flock to the healthy flock should be controlled. Crow and sparrow act as carrier in the transmission of the disease. Faecal excreta of more birds contain normally infectious organisms. These droppings get easily mixed in the drinking water stored outside the farm premises without protection. By this, many of the organisms spread the disease.

Prevention by vaccination is an important feature in the control of the diseases. Vaccination is a main factor for prevention of infection in infected flock. Planned vaccination is advisable. Vaccine dosage, vaccination technique and vaccination time determines the immune level. Normally the following vaccination schedule is recommended.

5th day - RDVF or LaSota by nasal drop or drinking water

8th day - RDVK subcutaneous in wing

In the possible places of outbreak the recommended schedule of vaccination is

5th day - RDVF or LaSota - Nasal or drinking water

4th week - RDVF or LaSota - Nasal or drinking water

8th week - RDVK by subcutaneous in wing

16th week - RDVK by subcutaneous in wing

Vaccination alone is not a criteria. Level of immunity as vaccine response have to be assessed. Vaccine potency, vaccination technique and vaccine response are taken into account for immunity level. In the unvaccinated flock, the out break induces 100 percent mortality.

During the outbreak, the immediate preventive precautionary vaccination is recommended. However it differs with age of the birds. In the adult grown up birds, revaccination with RDVK is necessary. Infected and incontact birds should be separated first. Afterwards individual vaccination by using separate needle is advisable to avoid transmission of infection by vaccination. Vaccine virus enters all the cells of the body. These viruses play a vital phenomenon of interference, wherein presence of vaccine virus in the cell prevents the entry of field virus, thereby control the infection. Now much higher immunity inducing killed vaccine with less stress is available. Owing to high cost, these types of vaccines are used in the breeder flock.

□□□

Chapter-9

Fowl Pox

In this disease, skin is the primary organ of attack. In addition, tissue lesions are seen as pustules in respiratory tract. In the egg yolk also the birds induces pock like lesions. In the affected birds, inclusion bodies are present in the tissues. The inclusion body is termed as bollinger bodies and particles present inside the inclusion bodies are known as borrel granules. In due course of multiplication of virus, necessary immunity is established.

Cause

Virus comes under the family pox viridae and is represented as Avipox. In this family, there are many viruses like turkey pox, pigeon pox, quail pox, sparrow pox. Chicken pox is a large virus group. Affected cell after death is get detached from the body. Such cells contain virus in alive condition. These cells are capable of infecting healthy cells easily.

Incidence

Organisms cannot enter the intact skin. For entry of the virus, injury or broken skin is necessary. Viruses spread through directly. Diseases can also be spread by other ectoparasites. Mosquito and flies are known to cause the infections. Infections through the respiratory tract is possible.

Symptoms

Viraemic condition is seen in blood. Entry and multiplication of organism into the body does not cause any damage. Organisms are multiplying in the cells and spread to the other places. During the course of spread, the virus induces the fact on for increasing the cell size. Due to these changes, the lesions are produced in the cells. Pock lesions are seen in head, neck, leg and foot region.

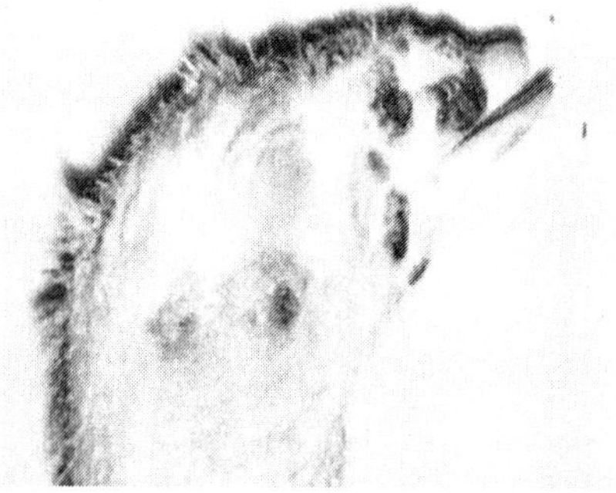

Cutaneous form (dry pox). Face lesions in young turkey

In the upper respiratory tract, diphtheritic ulcers are seen. In chicken pox, lesions are seen in skin. Onset of disease is slow. Incubation period varies from 4 to 14 days. In the initial stage, the pock lesions turn into pustules. In the course of the disease, the pustules get dried and fall as a dry scab. The scab contains lot of organisms which are infectious in nature. Lesions in the respiratory tract is acute in nature. It starts as small pustules and subsequently merged and appears as purulent nature. This type of lesion is seen in upper respiratory tract and digestive tract. Eyes affection leads to continuous lacrimation, inappetance and difficulty in breathing are quite evident. Mortality rate is 50%.

In layers it leads to drop in production. The course of the disease is for the period of 3 - 4 weeks. In complicated cases the affection and attack are more.

Diagnosis

- Symptoms
- Presence of inclusion body in the affected cells.
- Isolation of the organism and growing pattern in the embryonated egg.
- Detection of the antibody in the blood serum by serum neutralisation test

Differential Diagnosis

- Vitamin A deficiency needs to be differentiated from this condition.

Immunity

Recovered birds have solid immunity against disease.

Treatment

It is not possible to treat the affected birds. Affected birds can be treated symptomatically to reduce stress by giving anti-stress factors and also antibacterials to prevent secondary bacterial infections.

Prevention

Vaccines are available to prevent the diseases. Live attenuated vaccines are produced by growing in the eggs. Vaccines are to be given as two types. First vaccine needs to be given as wing web method, wherein broken head of the needle is used to charge the vaccine and puncture the wing web. After two weeks, the second fowl pox vaccine is given by the method of feather follicle in the thigh. Certain feather follicles in the thigh are removed and vaccine is given. Vaccine reaction in the vaccinated place is need to be seen. Inflammation and sore nature of the skin shows production of solid immunity. Lack of reaction in 10% of the flock requires revaccination. Initially less virulent vaccine of pigeon pox vaccine needs to be given at the age of three weeks. Subsequently fowl pox vaccine is recommended at the age of six weeks. Immunity produced by the pigeon pox vaccine is two months and fowl vaccine is much longer time. Now tissue cultured fowl pox vaccine as a single vaccination procedure is readily available. It is possible to introduce the antigens of the Ranikhet disease, Infectious bursal disease and Mareks disease in the non immune regions of the fowl pox virus affected parts.

❑❑❑

Chapter-10

Infectious Laryngeotrachitis

This disease is first identified during the year 1924 at America and subsequently in the year 1935 at England. Disease is prevalent in all places of Europe, Asia and throughout the world. Mortality and production drop depends upon the disease intensity.

Cause

Organism belongs to the family Herpes viridae. This virus belongs to the genus gallid herpes virus. There are different strains depending upon genetic basis with DNA structure. Intensity of disease and places varies with different parts. Organisms are alive in farm for many weeks. This virus grows in the embryo of chicken, duck and turkey.

Incidence

Chicken and peasants are attacked by this disease. This disease affects all age groups. Birds in the age group of 3 to 9 months are attacked frequently. Cocks and high weighing birds are highly susceptible. In view of the affection of the trachea, this disease is prevalent as a mixed infection with other respiratory infections.

Transmission

Transovarian infection is possible. Primarily disease transmission happens having contact with the discharges of the nose, mouth and cages of the infected birds. This discharge contains infectious organism in huge number. Organisms are spread by air. Infectious organisms enter the body through respiratory tract and eyes. Sometimes the organisms are localised and the organisms are excreted for many years. Stress is the major predisposing factor for the spread of the diseases. Live birds are responsible for transmission of disease. The organisms apart from live birds are capable of alive in farm equipments like feeders, waterers, equipments and building also spread the disease. Disease transmission is easy and happens by many ways.

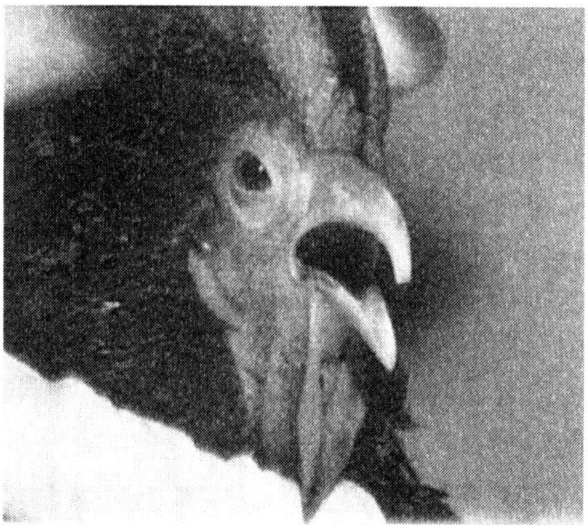

Difficult breathing and coughing

Incubation period for onset of disease is 6 to 12 days. This disease occurs as per acute, acute and symptomless. In per acute cases, there is a sudden mortality. Difficulty in respiration, cough and blood stained nasal discharge, blood clots are specific signs. Birds die within 1 to 3 days after the onset of symptoms. In acute cases, many symptoms are not seen; still there is a difficulty in respiration. Nasal block with mucopurulent discharges and sometimes blood clots are evident. Due to nasal block, the birds respires through mouth. During mouth respiration the beak is open. Owing to the difficulty in respiration, there is a sharp pitch noise. There is continuous discharge from nostrils. In the highly affected birds, eyes are swollen, inflamed reddish with lacrimation. After a period of 2 to 4 days of attack, the affected birds die. In the moderate and symptomless type, there is lesser intensity of breathing difficulty. Frequent shaking of head, nasal discharge and lacrimation are also evident in this type. Affected birds die on the ground. Egg production is usually affected.

Due to air borne transmission all the birds are affected easily. Depending upon the intensity of the disease, the mortality accounts to 70%. Period of attack in the farm is around two weeks and sometimes even for two months.

Lesions

Lesions vary with intensity of the disease. Major lesions are seen in the upper respiratory tract. In per acute cases, upper respiratory tract is affected completely. Hemorrhage is seen in the respiratory tract. Blood clots present in the trachea cause difficulty in respiration. Primary parts of the bronchi are also affected. In acute cases, the intensity of similar lesion in respiratory tract is moderate in nature. Moderate hemorrhage, petechiae, purulent pustules and thick mucus discharges are present in acute cases. Inflammation of the respiratory tract leads to reddish and sometimes cyanosed colour. In the moderate and symptomless type, the lesions are lesser. Purulent pustules and thick mucous discharge, blocking nasal tract are seen. Eyes are inflamed. Inflammation of lungs is present in all types of infection. Disease should be differentiated from fowl pox and nutritional croup due to vitamin A deficiency.

Presence of inclusion bodies in the affected cells of the epithelium is the confirmation steps.

Diagnosis

It is easy to spot the disease by seeing the specific symptoms and lesions. Isolation and identification of the causative organism from the infected birds is the confirmative one. Presence of antigen in the blood serum is confirmed by the serological test. Histopathological examination of the affected cell shows presence of intranuclear inclusion bodies.

Control Measures

Farm hygiene and good environment is the primary factor in control of the disease. Use of right type of disinfectant in the cleaning farm house and farm equipment is advisable. Infected flock should not be allowed to come into contact with healthy birds at any place. Rats and dogs movements should be avoided.

It is possible to control the disease by vaccination. Vaccine needs to be used in outbreak places alone. In the disease free zone, vaccine should not be used, since some birds continue to excrete the virus, mortality imitates the disease out break in healthy birds. Avirulant vaccine can be used safely. Vaccination can be done at the age of 3 to 10 weeks by nose, feather follicle, drinking water and cloaca. Booster vaccination is necessary after 8 weeks.

❑❑❑

Chapter-11

Infectious Bronchitis

Infectious bronchitis is caused by RNA virus named Corona virus. Affected birds manifest mostly respiratory symptoms. In layer birds drop in production is seen. The chicks die due to involvement of kindey and lungs.

Massachutes and Connecticut are the two classical types of diseases which affects the lungs. Infectious organisms are also seen in small intestine, large intestine and caecum region. Virulence of infectious organisms is changed during the growth in the embryo. It is easy to eliminate the organism using disinfectant. At a temperature of 5°C for the duration of 30 minutes , the organism losses it virulence.

Incidence

Naturally the disease occurs in chicken. The disease mainly affects the reproductive tract of the chicken. In the chicks, it affects the lungs and thereby causes great damage. Kidneys are affected in the young aged birds. In the poultry, depending upon the resistance of the birds, intensity of the attack varies. In the layers due to changes in hormone profile during the onset of laying is accounted for excretion of the infectious organisms. In the mixed infections with infectious Laryngeotracheitis, Ranikhet disease, E-coli, and Mycoplasmosis, the intensity of attack is more. Cold stress is one of the predisposing factors for this disease.

Transmission

Organisms are localised in the birds for many months. In this way, it is possible to transmit the disease from one bird to other. Air borne transmission is very much feasible from infected birds to the healthy birds. Droppings of the birds also contain organisms. Vertical transmission by way of trans-ovarian infection is possible. Infectious organisms also spreads in various equipments used in the farm. Organism are sometimes get localized in the birds without any affections. Incubation period is 1 to 3 days. Air borne transmission requires 24 hours for onset of disease. Natural cases require just 3 hours to induce the disease.

Signs

Signs differ depending upon place of affection either it is in lungs or in reproductive tract. Generally, poor retarded growth, off feed, poor feed efficiency, dullness, ruffled feathers, are equally important signs. Birds affected with bronchi exhibits noise in breathing, cough, nasal discharge, lacrimation and swelling of face. Symptoms are clearly seen in all the birds. Disease is rapidly transmissible. Mortality rate if affected with other disease extends up to 30%. In the respiratory affection much mortality may not be there. In the mixed respiratory infection, the duration of affection ranges between 10 to 14 days. Infected broiler birds affected with other diseases results in higher level of mortality.

Chick – Respiratory signs

Misshapen eggs

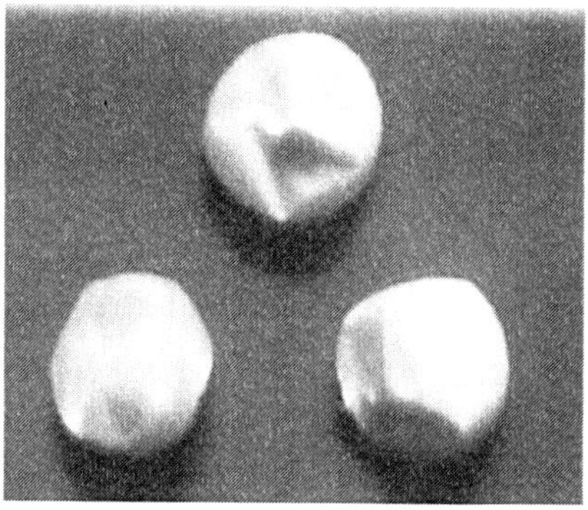

Shell-less egg

This disease affects the reproductive tract by two ways. In the first type incidence occurs more in layer birds and in the second type chicks are affected. In the adult layer birds, ovary and oviduct are also affected. Infection during the period of peak egg production reduces the egg production to the level of 50%. It requires 4 to 5 weeks to reach previous egg production. Even though the egg production could be restored, the quality of egg is infected continuously. Reduced egg size, misshapen, rough ridges in egg surface and discoloration of egg and conspicuous feathers are seen. Egg white is watery; collagen of the egg is broken leading to floating of the egg yolk. Hemorrhages are seen in the egg white and egg yolk. Intensity of affection depends on the immunity of the birds.

In the second type, reproductive tract in chicks are poorly developed. In these birds, released egg yolk cannot be enter into the reproductive tract, but it drops in the stomach. These types of birds are called as blind eggers. Sometimes eggs are developed, however quality of the egg is poor.

Chicks in the age group of 3 to 6 weeks, kidneys are affected. Dullness, breathing problem are mainly seen in the affected chicks. Mortality may extend up to 80%. In the avirulent type there is no such mortality.

In the new emerging type, affection shows moderate difficulty in respiration, 5 to10% drop in production and produce colourless egg yolk. It requires 2 to 3 days to bring the egg production to the normal level.

Lesions

Lesions vary according to the size of affections. In the respiratory type, lesions are seen in the lungs, bronchi and nasal tract. Inflammation of the lungs and enlargement of air sacs are evident. In the acute type, nasal discharge gets thickened and blocks the nasal tract. Inflammation of the trachea with petechiae is quite evident. In the microscopical examination of mucous membrane of respiratory tract, the cilia are affected with enlargement of the epithelial cells.

In the reproductive tract, manifestation of the uterus is developed with changes in the shell structure. In the chicks, the uterus is under developed.

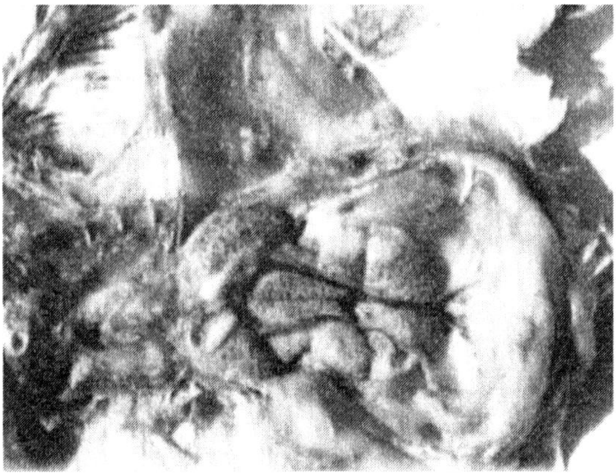

Nephritis – Broiler chicken

In the kidney affection, enlargement of kidney and blockage in ureters are visible and contains excreted products.

Some times lesions are also seen in the intestine. In the layers, the eggs laid in the abdomen are decomposed with foul smelling.

Diagnosis

Symptoms and lesions are the important criteria for diagnosis. Differential diagnosis with the diseases affecting kidney and lungs are quite important criteria. Isolation and identification of the causative organisms is another important method to show presence of antigen in the blood sera by laboratory investigation. Organisms can be grown in embryonated eggs for confirmation. ELISA techniques are now used for identification of the disease.

Control

Farm hygiene and management procedure must be in satisfactory level. There should not be any difficulty in ventilation. Vaccine is necessary to prevent the disease. Live and killed vaccine is used. Live vaccine have to be given eight weeks before the killed vaccine.

□□□

Chapter-12

Infectious Bursal Disease (Gumboro Disease)

Bursa of Fabricious responsible for humoral immunity gets affected in this disease. Birds more than three weeks of age are usually affected. This disease was first identified in the place of Gumboro, hence known as Gumboro disease. As it affects the kidneys, it is also termed as Avian Nephrosis. In India, the disease is found in Uttar Pradesh during the year 1971. In Tamil Nadu, incidence was reported in the year 1987.

Cause

The virus comes under the family Birnaviridae. There are two types of antigen in the virus and hence Virus is differentiated as two types. In the first type organism affects B-lymphocyte cells which leads to drop in immunity level. This disease comes under acute to moderate type. The second type does not cause the disease. Both the types affect turkey and ducks.

Infectious organisms lose viability at a temperature of 70°C for 30 minutes. It also losses infectivity by exposure to 0.5% formalin, 0.5% chloroquine for the time of 10 minutes destroys the infectious organisms. As the infectious organism is harder, it is difficult to clean the farm using the disinfectant.

Incidence

Type1 Virus affects the chicken, turkey and ducks very seriously. Infectious birds are continued to excrete organism from 10 to 14 days. In the farm, infectious organisms in surrounding environment alive for longer periods with good infectivity. Mites and worms in the litter transmit the disease from one place to another. Chicks with good maternal antibody are able to withstand the disease. Drop in egg production leads to infection. White leghorn birds are affected early. In the grown up birds, the bursa naturally weakened and the incidence of infection is less.

It takes 2 to 3 days to induce infection in birds. This organism affects lymphoid cells present in bursa of Fabricius, liver and caecal tonsil. T lymphocytes are not destroyed. Generally, the infectious organism enters body through respiratory tract and eyes. Infectious organism reaches internal organ of the body within six hours, intestines within 12 hours and bursa of Fabricius within a hour. In the bursa of Fabricius antigen will be present for 3 days.

Enlargement of kidney with blockage in the ureter is one of the affection. In the mucus there is a frank haemorrhage.

Infectious bursal disease depends on type and virulence of the organisms. The course of disease reflects on how much the immune system is affected.

Immuno-Suppression Type

In the birds, the disease affects the young ages. Bursa of Fabricius is completely affected. The chick gets infection at the age of 3 days, resulting in complete cessation of physiological function of bursa of Fabricius. In 7 days old chicks, 30 % of the bursa is affected. Chicks of 15 days age show lesser damage in the bursa.

This disease affects the chicks of 2 to 3 weeks age. Mixed infection with other disease causes much mortality. It also affects six weeks birds.

Moderate Type

Usually affects the chicks of six weeks. Disease in aged birds as such there is no damage. Even then the immunosuperession is temporary. After attack, the birds revive to normal position.

Acute Type

Depending upon the type, this disease is classified. During the out break, the mortality rate is reached up to 60 to 100%.

Different Types

They are identified in United States of America. It differs from other types. Antigen variety is different in nature. Infectious rate is also different. More importantly there is a depression of immunity in birds is also seen. Even with good immunity, the outbreak happens.

Signs

Signs differ with virulence of the organism. Acute type of disease is seen in 3 to 5 months birds. Dullness in birds, off feed, watery diarrhoea, dropping in cloaca, inappetance, ruffled feathers, eyes closed leading to mortality are common signs. Disease transmission rate accounts to 10 to 100%. Mortality rate is 20%. Sometimes it records 50%.

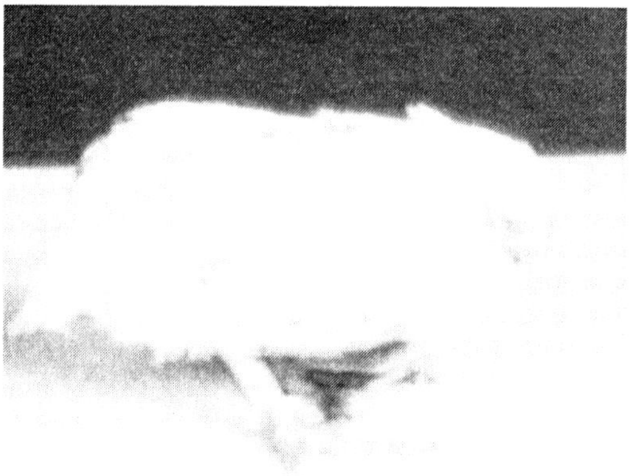

In the moderate type, symptoms are fewer. Poor growth and breakdown of immunity to other diseases are seen. The course of disease ranges from 3 to 5 days. Depends upon the immune level, the course of disease may occur.

Lesions

Dead birds are in debilitated condition. Petechial hemorrhages in thigh and legs are seen; liver and mucosa of intestines also contain petechial hemorrhages. Enlargement and inflammation of the liver and spleen are quite evident. Bursa is Inflammed and edematous. In the period of 3 to 8 days, the bursal organ losses physiological function and becomes smaller in size. Caecal tonsil contains hemorrhage and thick caseated mass.

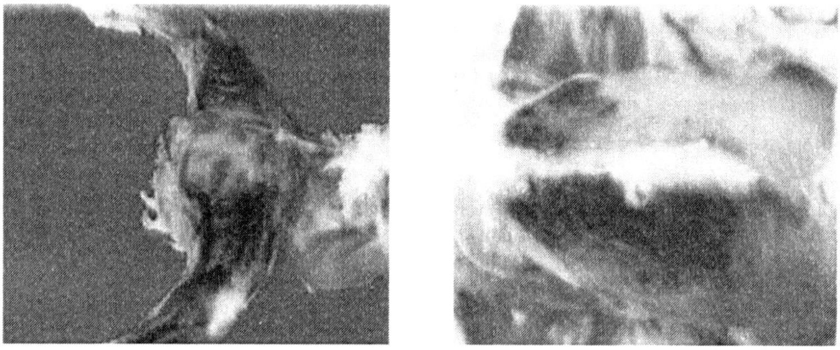

Hemorrhages in the thigh and pectoral muscles

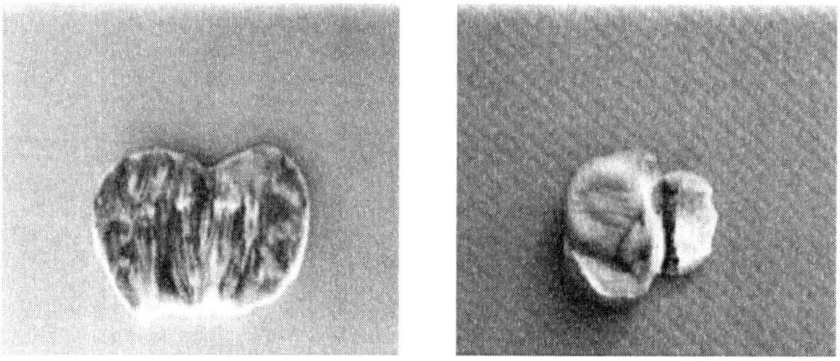

Haemorrhages in the Bursa of Fabricius and Proventriculus-gizzard junction

Kidneys are swollen and pale. Ureter is enlarged due to accumulation of excreted products. Variation of lesion in kidneys is seen.

Diagnosis

Initiation of disease, symptoms and lesions are primary factors for diagnosis. It is necessary to differentiate from other diseases. Antigen and antibody reaction is considered to be the confirmative diagnosis.

Prevention

It is difficult to destroy the organism. Infected place should be thoroughly cleaned using some disinfectant. In the infected place, all the birds should be disposed and should be kept the farm empty for sometime. After depopulation, strict sanitation measures are necessary before restraining the farm.

Maternal antibodies are quite useful to prevent the disease in chicks. As to establish maternal immunity, the breeder flock should be vaccinated at the age of 4 to 10 weeks with live vaccine and subsequently at 16 weeks with killed vaccine. Maternal immunity persists for three weeks only in chicks. At 30 weeks medium virulent vaccine could be used for direct immunity. Even then the immune levels vary in condition. Sometimes this vaccine virus can induce outbreak. In this place of out break field virus should be used to vaccinate parent stock. Chicks are need to be vaccinated with one type vaccine. During the time of less or lack of maternal immunity, if the chicks are vaccinated with medium type virulent vaccine, it may result in immuno depression or sometimes causes out break. As such it is advisable to use moderate virulent vaccine first at the age of 2 weeks and then with medium virulent vaccine as second vaccine.

□□□

Chapter-13

Mareks Disease

It is one of the most important diseases in poultry. This disease affects the lymphocytes of the white blood cells. This disease was first identified by Joseph Mareks at Hungary during the year 1901. As it affects nerve, there is paralysis. Lymphocytes nodules are produced in all internal organs. In the year 1961, the mareks disease is differentiated from lymphoid leucosis.

Cause

Lipotrophic herpes virus is responsible for cause of the disease. There are three types of disease. First serotype causes tumourous nodules. Second serotype is non pathogenic in nature. But it is identified that it also causes disease. Third type is non virulent and to be non pathogenic in nature. Pathogenic serotype is further classified as less virulent, virulent and more virulent. This virus combines with cells. However in feather follicle cells, it is present as separate cells.

Transmission

This disease is seen throughout the world Management and environment varies with the intensity and age of disease attack.

Affected birds continue to excrete the infected organisms throughout its life time. Infectious organisms are present in feather follicle cells, droppings and nasal discharges. Organisms present in feather cells alive for period of one year. There are two types of genetic disease resistance remain in birds. One is present in major histocompatibilty complex and another at cellular level. Cocks are highly susceptible than layers. Stress in the birds is responsible for cause of the disease.

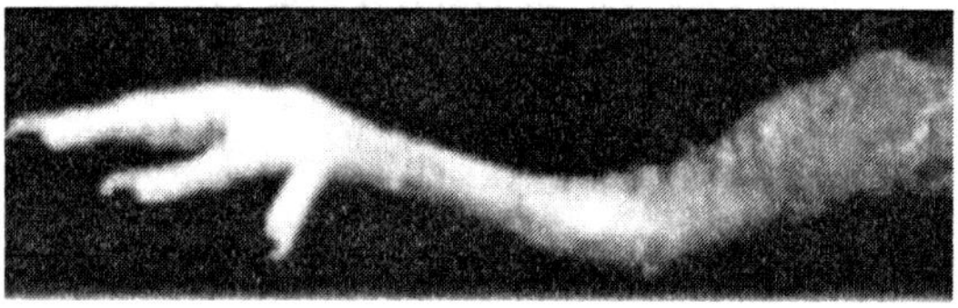

Marek's disease – skin involvement

Symptoms

Marek's disease affects certain age group of birds. Disease attack is seen in birds of 5 to 24 weeks of age. This disease is appeared in two types, as ordinary and acute type.

Unilateral paralysis of legs

Mortality rate in affected flock is 10 to 15%. Mortality of birds is seen for few weeks and sometimes for some months. This disease affects nervous system. Due to involvement of nervous system, there is paralysis of legs and wings. In some birds, nerves in the neck also gets affected, leads to torticolis. Paralysis and subsequent off feed results in mortality.

Acute Type

Percentage of mortality is high and in the range of 30 to 80%. Affected birds die from few weeks to some months continuously. Sudden mortality of the birds is also seen. Paralytic and depressed birds succumb to disease.

Lesions

In the ordinary type, affected nerves are enlarged; cement coloured and edematous, differentiating easily from the healthy nerve. More frequently, sciatic, brachial, coeliac and vagus nerves are affected. In these places, tumorous lymphoma nodules are seen. In this type, tumors are seen as cement coloured and soft in texture. Tumour growth is also seen in the lungs, heart and stomach. Enlargement of liver differ from one bird to another.

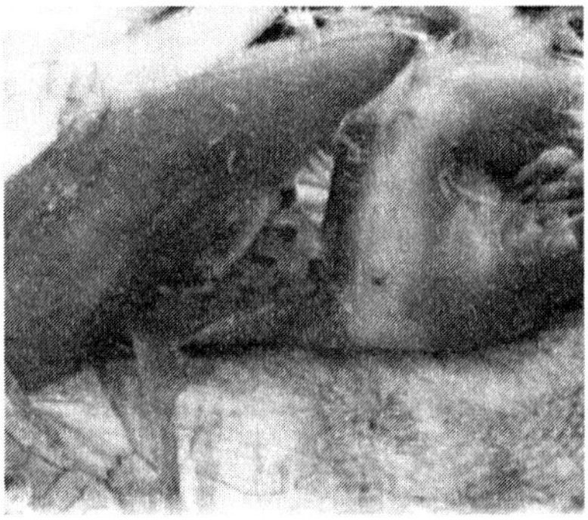

Subcutaneous hemorrhage

Tumours –Liver, heart and spleen

Tumour-Ovary

In both types, lymphoid type of white blood cells is encountered. In acute type, the lymphoid cell count increases and expressed as lymphoid nodules.

Pathogenicity

Infectious organism enters the lungs and multiplies in non lymphoid type of cells. In acute type, the disease manifests in 4 to 5 days. Organisms destroy the cytoplasm in the cells. Toxigenic virus and other ungrown virus grow in the B lymphocytes. Lymphocyte virus enters the body and attacks the nerves.

Diagnosis

Signs, lesions and presence of specific antigen in the blood are the primary factors for diagnosis. Isolation of the organism in the feather is necessary for confirmation. Mareks disease needs to be differentiated from lymphoid leucosis. The following points are used to differentiate these two diseases.

Criteria	Mareks Disease	Lymphoid Leucosis
Age	6 to 24 weeks	Above 20 weeks
Signs	Paralysis of legs	Drop in weight gain
Affection	Not more than 30%	Not exceeding 5%
Lesions		
Nerves	Present	Absent
Bursa of Fabricius	Poor growth or enlargement	Presence of tumor
Liver	Moderate Enlargement with tumerous nodules	Severe enlargement with tumerous nodules
Cell affected	T-cells	B-cells

Preventive Measures

Preventive measures depend on management factors and genetic disease resistance. In the management condition, the birds should not be allowed to get the infection. Birds in the age group of first four months needs to be grown carefully, without giving any chances of contacting the disease. Genetic disease resistance needs to be given due attention in the flock.

Day old chicks should be vaccinated at day old itself. Immunity is established within a week after vaccination.

This vaccination method is tried for mareks disease, wherein the vaccine is administrated on 18th day incubated egg.

❑❑❑

Chapter-14

Lymphoid Leucosis

Leucosis sarcoma causes tumour condition in the red blood cells, white blood cells, muscle, kidney, bone, liver and spleen. Lymphoid leucosis is a condition affecting layers. This disease also manifest in layer birds with drop in production alone but without any clinical signs.

Cause

Tumour causing virus affects the birds and establish the disease. This virus comes under the family Retroviridae. This tumor causing virus is classified as A, B, C, D, E and J & the viruses are classified by studying the glycoprotein content of the virus.

Molecular genome of this virus present in chicken and other birds at full or partial level. This genome is seen in the somatic and reproductive cells. Such joint cells are called as 'endosome viral loci'. These cells reproduce as per mendelian law. Endogenous virus varieties are known as E type. Virus infecting from external sources are remarked as A, B, C, D, and J types Tumour causing property is less in E type. Each type of virus causes separate type of tumour.

Sign

Anaemic and shrinked comb and wattles, dull, depressed and off fed are some of the signs. Egg production, egg size, hatchability rate, fertility rate and growth rate are dropped significantly. Breast bone bend with less muscle texture. In the case of liver tumor, enlargement of liver is observed. Enlargement of bursa is also evident. Bones are also affected. In the layer, egg production is affected. Affected birds die ultimately.

Pathogenicity

Lymphoid leucosis is one of the commonest diseases affecting the liver. Increase in number of lymphocytes are noticed. Tumour nodules are enlarged. Tumour nodules are seen inside of the livers and are smaller in size. Tumours present in internal organs of the birds. Tumours also develop in the bursa of fabricius. In the tumour producing organs, immature lymphoblast cells are quite evident.

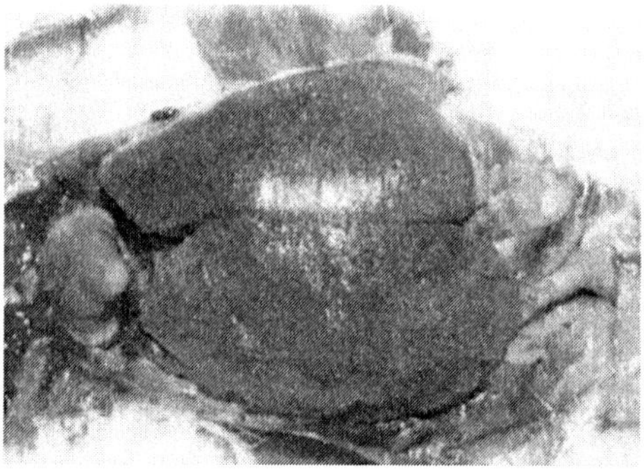

Incubation period extends to four months. Tumour producing cells are also seen in bursa.

Erythroid leucosis causes tumour conditions in white blood cells especially in bones. In this type, liver, spleen and inside of the bones are reddish in colour. Enlargement of liver and spleen is seen. Myeloid leucosis condition affects bones apart from this. Other type tumour in muscle, skin; red blood cells and liver are also seen. This leucosis is very common in all parts of the world.

Transmission

This disease is transmissible from parent stock to the day old chicks through egg. Infected birds spread this disease directly and indirectly.

In the egg transmission, viruses are present in the albumen. In the infected parent stock, the organisms are continuously disseminated through reproductive tract.

Infectious organisms are present in saliva, faecal droppings and feather follicles; virus is easily destroyed in the surrounding environment immunity.

Parental antibodies through egg yolk established immunity in the chicks for 3 to 4 weeks. Affected birds are capable of preventing multiplication of virus by its humoral and cellular immunity.

Genetic disease resistance is established by two ways.

1. By prevention of infection
2. By preventing the tumour production

Diagnosis

Virus character and pathogenicity are basic steps taken to identify the type of leucosis. Microscopical examination of the affected tissue depends on isolation and identification of the causative organisms from infected birds, red and white blood cells, liver, tissue, reproductive discharge, albumen, and egg yolk. Antigen and antibody reaction are also used to confirm the disease.

Prevention

Chicks should be purchased from parent stock free from this disease. Farm environment should be kept in hygienic condition. Breeder flock of good genetic resistance is advisable to use right type of disinfectant to eliminate the disease organisms.

□□□

Chapter-15

Avian Influenza

Synonyms

Bird flu, Avian flu, Influenza A flu, Type A flu or Genus A flu, Food Plague.

Introduction

Avian influenza or bird flu is a contagious disease caused by viruses that normally infect only birds and less commonly, pigs. Avian influenza viruses are highly species-specific, but have, on rare occasions, crossed the species barrier to infect humans, hence it is zoonotic. The disease was identified in Italy (around 1878) more than 100 years ago and now occurs worldwide.

Etiology

Family: orthomyxoviridae

Genus: Type A Influenza virus

Species: Avian Influenza virus

It is a single stranded negative sense RNA virus and the genome is arranged in 8 segments. The outer surface of the virus particle consists of a lipid envelope from which projects prominent glycoprotein spikes which are of two types, Haemagglutinin (HA) and neuraminidase (NA).

Avian variants

Low pathogenic Avian Influenza (LPAI)	Highly Pathogenic Avian Influenza (HPAI)
Little or no mortality	Called as fowl plague, Deadly flu or Death flu

OIE Listing

List A disease.

Resistance

Avian influenza viruses are not very stable and are relatively sensitive to detergents. They are destroyed by formalin, oxidizing agents, dilute acids, heat, extremes of p^H and dryness. But the influenza virus can survive for long period of time in the environment, particularly under cold and most conditions.

Haemagglutination Properties

Avian influenza virus causes agglutination of chicken erythrocytes.

Viral Subtypes

Influenza A viruses are classified into subtypes on the basis of their Haemagglutinin (HA referred as H) and Neuraminidase (NA referred as N). There are 16 different HA antigens (H1 to H16) and nine different NA antigens (N1 to N9) for influenza A. Until recently, 15 HA types has been recognized, but a new type (H16) was isolated from black - headed gulls caught in Sweden and Netherlands in 1999 and reported in literature in 2005. There are three subtypes of avian influenza virus referred as H5, H7 and H9, of which H5 and H7 can produce low pathogenic and highly pathogenic avian influenza, where as H9 produces only low pathogenic infection. Within these subtypes there are number of strains. These strains are formed by combination of any one of H5, H7 or H9 with 9 different NA proteins.

Sub. Type	Characters	Example
H5	Nine different strains	H5N1
	Produce highly pathogenic or low pathogenic infections	
	Human infection possible, sometimes causing severe illness and death	
	Nine different strains	
	Produce highly pathogenic or low pathogenic infections	
H7	Human infection is rare, but can occur among persons who have close contact with infected birds ,symptoms may include conjunctivitis and / or upper respiratory system.	H7N7 (Netherland)
H9	Nine different strains produce only low pathogenic infection	H9N2
	Human infections possible.	

The avian influenza virus currently caused the major epidemics in Asia is H5N1, with some occurrences of H5N2 and H9N2.

Hosts Affected

Avian influenza viruses are distributed throughout the world in many domestic birds including Turkeys, chickens, Guinea fowl, Quail, Pheasants, Geese and Ducks and in many wild species. Wild fowl acts as natural asymptomatic carriers of influenza A virus.

Transmission

Water fowl serves as reservoir for AIV and responsible for introducing the virus into poultry virus which is found in large quantities in faeces, respiratory and conjunctival secretions of infected birds.

- Air borne /inhalation
- Oral route.

It also spreads through contaminated equipments, egg flats, feed trucks, shared contaminated drinking water etc. No vector transmission.

Incubation Period

3-7 days depending on the subtypes, age and species of the bird.

Morbidity and Mortality

Morbidity and Mortality varies according to species and virus. Incase of highly pathogenic viruses the morbidity and mortality reaches upto 100%.

Pathogenecity

Influenza viruses bud from the apical surfaces of polarized epithelial cells (Eg:- Bronchial epithelial cells) into the lumen of lungs and are therefore usually pneumotrophic. In same cases, the disease is rapidly fulminating and birds found dead without any signs. In highly pathogenic form, there is highly significant decline in feed and water consumption and egg production, respiratory signs were less prominent but the birds were severely depressed and some had tremors or unusual attitudes of head. "Pigs = Mixing Vessels".

Clinical Signs

The clinical sign varies according to species, age, concurrent viral or bacterial disease and environment.

Infection with Non pathogenic viruses

- No clinical signs, with seroconversion
- Some of these viruses have the potential to become virulent through genetic mutation.

Infection with low or mild pathogenic viruses

- In apparent to mild or severe respiratory disease
- Mortality ranges 3% in layers and 15% in broilers
- Drop in egg production
- Mutation to virulence has been demonstrated

Infection with highly pathogenic viruses

- In per acute cases, sudden death occurs.
- In acute cases, mortalities occur as early as 24hours after the first sign and frequently death within 48hours.
- With excessively watery eyes and sinusitis, cyanosis of the combs, wattle & shanks, edema of the head and eyelids, ruffled feathers, diarrhoea and nervous signs. Eggs laid after the onset of illness frequently have no shells.

Lesions

In per acute cases, there is lack of visible gross lesions other than severe congestion of musculature and dehydration. In acute cases, there may be subcutaneous edema of the head and neck area, exudation of fluid from nares and oral cavity, congested conjunctiva with petechiation, tracheal lumen filled with excessive mucus exudation, hemorrhagic tracheitis, pinpoint hemorrhage on the keel, very small petechiae over the abdominal fat, serosal surfaces and peritoneum and congestion of kidney with urate deposition in the tubules. Cyanotic Comb and wattles present. As the disease progresses, the pancreas, liver, spleen, kidney and lungs can display yellowish necrotic foci. The peritoneal cavity is frequently filled with yolk from ruptured ova. Hemorrhages may be present in the proventriculus particularly at the junction with the ventriculus (gizzard).

Diagnosis

The disease can be diagnosed by symptoms, isolation and identification of organisms by cultivation in embryonated chicken egg, RT-PCR and serological tests like AGID, HA/HI, ELISA.

Differential Diagnosis

From diseases causing high mortality

- New Castle disease
- Infectious laryngeo tracheitis
- Acute poisoning
- Diseases causing swelling of the combs and wattles
- Acute fowl cholera and other septicemia diseases
- Bacterial cellulitis of the comb and wattles.

Treatment

Practically there is no specific treatment for avian influenza viral infections. Amantidine Hydrochloride is effective against infection of quails, Turkey and chicken.

Control

Inactivated oil emulsion vaccines are found effective in chickens and turkeys. Avirulent or attenuated strains can be used as vaccines. A recombinant fowl pox virus vaccine containing the gene which codes for the production of H5 antigen can be used. The problem with live vaccine is emergence of new strains due to genetic reassortment.

Avoiding free range birds in waterfowls frequent areas. Proper biosecurity programmes of proper cleaning, quarantine of birds and unknown disease into the flock and disinfection of farm premises and equipments.

❑❑❑

Chapter-16

Avian Encephalomyelitis

Introduction

Avian encephalomyelitis is an infectious viral disease and it affects the chickens, pheasants, turkeys and quail. It has a world wide distribution. It is characterized by ataxia and rapid tremors, especially of the head and neck; because of the latter, it was often called as "epidemic tremor". Morbidity 50-60%; mortality is high. It is of economic importance, due to the infection it causes reduced egg production in laying hens and a lowered hatchability of fertile eggs. Vertical transmission also occurs. Chicks of about 1 to 2 weeks of age are usually affected. The incubation period is one week. Virus in faeces may survive for more than 4 weeks.

Transmission

Virus spread mainly by Vertical transmission. Horizontal transmission also occurs through the infected birds droppings. Infection also occurs through fomites.

Clinical Signs

Clinical signs include dull, depression, ataxia and sitting on hocks. Nervous signs like paralysis, imbalance, tremor of head, neck and wings. Tremor may be in apparent but is accentuated if chicks are held inverted in the hand. Tremor may be absent in some outbreaks (or) may occur only in few birds. Ataxia usually progresses until the chick is incapable of moving about and this stage is followed by inanition, prostration and finally death. In layers, the drop in egg production is about 5 to 10%.

Lesions

Gross lesions are mild (or) absent. There may be focal white areas in gizzard muscle. A few recovered birds may develop cataracts, weaks after infection. Microscopically non purulent diffuse encephalomyelitis with perivascular cuffing is seen. No changes have been described for infected adult birds.

Diagnosis

Diagnosis is based on history and signs. Serological tests include immuno fluorescent assay, ELISA and laboratory tests include histopathological examination. In differential diagnosis this disease is differentiated from Newcastle disease, Marek's disease, vitamin deficiency diseases, nutritional encephalomalacia and mycotic encephalitis.

Control

This disease cannot be prevented by using standards of hygienic procedure alone. Control of Avian encephalomyelitis mainly depends on vaccination. Both live and Killed Vaccines are also available. They are giving long lasting immunity. But the live vaccine usually causes a drop in egg production if it is given to laying birds.

□□□

Chapter-17

Avian Reovirus

Introduction

Reoviruses are double stranded RNA viruses. Reovirus infections are prevalent world wide in chickens and turkeys. Reoviruses are commonly found in the digestive and respiratory tracts of clinically affected normal chickens and turkeys. Viral arthritis / tenosynovitis are found primarily in meat-type chickens and turkeys.

Transmission

- Horizontal transmission
  1. Intestinal tract (fecal contamination).
  2. Respiratory tract.
  3. Age related resistance.

- Vertical transmission
  1. Egg transmission.
  2. Hens infected via oral, tracheal and nasal inoculation is able to transmit reovirus to their progeny.

Clinical signs / Lesions

- Viral arthritis / tenosynovitis
- Lameness
- Joint swelling
- Thickened / ruptured
- Malabsorption syndrome
- Reverting / stunting
- Poor Pigmentation
- Abnormal feathering
- Skeletal abnormalities
- Increased mortality
- Enlarged proventriculus

Pathogenicity

Reoviruses have been identified as the ideology of other disease condition such as Arthritis / tenosynovitis, runting, pericarditis / myocarditis, hydro pericardium, hepatitis, Bursa and thymus atrophy.

Diagnosis

- Diagnosis is made on the basis of signs and lesions.
- Virus isolation in chicken embryo liver cells (or) FA test on infected tendon sheaths.
- Histopathology of infected tendons.
- Agar-gel precipitation test, Immunofluoresence assay and ELISA test helps to diagnose the disease.

Control

- Good husbandry and bio-security practices will help to reduce the spread of reovirus from flock to flock.
- Vaccines are used to prevent egg transmission to progeny and also produce maternal antibodies for the progeny.

❑❑❑

Chapter-18

Egg Drop Syndrome

Adeno virus causes this disease. The virus is commonly seen in goose and ducks.

Transmission

The disease spread by many ways. Vaccines are used in the breeder stock. Vaccination programme sometimes contain the virus. Transovarian transmission from parental stock to day old chicks through the egg is possible. During the time of peak egg production, the virus attains the virulence and spreads the disease. The virus inhabits ducks and flying birds and causes infection in all age groups and in all seasons. Horizontal transmission from the infected flock to the healthy birds is also seen. Mechanical transmission through egg trays is also possible. Secretions of the reproductive tract contain the virus and hence the droppings also contain the infectious organisms.

Signs

Important signs are drop in egg production and poor quality eggs. Thin shell egg and shell less egg is common feature. Mineral deposition, striations and variation in the egg shell are the characteristic changes. Affected birds will appear normal. Sometime off feed and drop in feed consumption are noticed. Faecal dropping contains water. Secretions in the reproductive tract of the layer birds are observed.

Misformed and soft shelled eggs

In the individual type, there is drop in production. Drop in production is recorded at the peak of the egg production. Expected level of peak egg production is not there. Antibodies may or may not be present in the blood. Egg drop syndrome is also seen in the quails. Drop in egg production and thin shell eggs are visible in the quails.

Diagnosis

Important signs like drop in egg production and egg shell affections are used to spot the disease. It is difficult to identify the disease in the cage layer birds. Disease transmission sometimes get delayed and occasionally within few months period. Affected birds are difficult to identify. Affected eggs can be given as a feed to the healthy birds and observing for the onset of the disease. In this way the disease occurrence can be confirmed. Isolation of the causative organism from the reproductive tract of the birds is an important laboratory procedure.

Presence of the antibody in the blood is used for confirming the disease. ELISA technique is used to assess the antibody level in the blood.

Preventive measures

Known disease free breeding flock is a basic approach. By this way it is possible to control the vertical transmission of the disease.

Attenuated vaccine can be used to establish immunity in the susceptible flock. This vaccine needs to be given between the egg groups of 14 to 18 weeks. This vaccine can be given along with bursal disease vaccine and Ranikhet disease vaccine. Disease transmission through eggs and egg trays should be avoided.

Disease transmission from flying birds like ducks and goose are to be controlled.

□□□

Chapter-19

Chicken Anaemia

This disease is caused by adeno virus. In this disease, there is anaemia, drop in lymphocytes, liver changes and haemorrhages in all parts of body are seen. This disease was first seen in Japan during the year 1979. This disease is prevalent in all parts of the world.

Adeno type of virus causes infection. In the affected birds the bone marrow, the site of red blood cells production in chicken are severely affected. There is haemorrhage in the places of immune organ. This disease is transmitted both vertically and horizontally. Breeder flock transmits the disease to the chicks. In egg transmission, the infection set in all the age within two weeks without any symptoms. Infected chicks also transmit the disease to the healthy flock. In both transmissions, there are heavy damages.

Chicken anaemia causing virus is not susceptible to the common disinfectant. Iodine and formalin are not useful at all to destroy the infectious organisms. Hypochlorite type of chemicals is the most appropriate disinfectant to kill the organisms.

Transmission

Disease is transmitted by direct and indirect means. In the act of egg transmission, other means of mechanical transmission by way of droppings, feeders, waterers and growing place also happens. Infections in the breeder flock does not show any clinical signs like drop in egg production and hatchability in the parent stock but the virus get activated and become virulent. Then this virus reaches the chicks. Even after hatching, the incidence of attack may also be there. However symptoms are evident after two weeks. Parent layer stocks continues to excrete the infectious organisms for a period of six weeks.

The important act of development of virulence happens in the chicken is used for vaccine production. As such, eggs produced from this flock also contains virus. These eggs if it is used for vaccine production necessarily incorporate the infectious virus. Hence by process of vaccination, the disease gets spread.

The intensity of disease varies by many factors. In the affected birds, it simultaneously occurs with Mareks disease and Bursal disease. In reticulo endothelial system, the intensity of attack is in severe condition. All these disease in combination attack all birds in all age groups within current immuno depression resulting in severe attack.

Symptoms

Symptoms vary with nature of spread. In the egg transmission, symptoms are seen at the age of two weeks only. Mortality rate ranges from 10 to 60%. In anaemia, red blood cell production in bone marrow is affected. Defense organ like thymus, spleen and bursa of Fabricius are atrophied. Haemorrhages are seen in muscle and skin regions. Inflammation and necrosis of the skin are present. Enlargement of liver is seen. Paleness of the wattles and comb and feather droppings are observed. Affected birds are dull and off feed. The incidence of this disease is seen in layer birds. Owing to drop in blood producing cells and T lymphocytes, the anaemic condition and immuno depression condition are noticed.

Diagnosis

Important symptoms are useful to diagnose the affected flock. Isolation and culturing of the causative organisms are used for confirmation. Laboratory investigations like ELISA and serum neutralisation tests are used to specify the disease.

Control

Disease free parent stock is very essential. Preventive vaccination is advisable in breeder flock and in chicks to establish the solid immunity against the disease. Live vaccines are used in the chicks. By any means, any other vaccines should contain the infectious agent of chicken anaemia.

□□□

Chapter-20

Inclusion Body Hepatitis

Introduction

Inclusion Body Hepatitis (IBH) was first described in U.S. in 1963 and then reported in many countries around the world subsequently, it was determined to be associated with avian adenovirus (AAV) infection. The first AAV associated with clinical disease was isolated from an outbreak of respiratory disease in quail in 1950. Since that time, AAV have been found in all types and breeds of chickens (normal flora). They have been isolated from numerous species of birds, including both healthy and sick poultry. In addition to IBH, AAV has been associated with hydropericardial syndrome (HP), egg production drops, reduced growth rate and feed conversion, tenosynovitis, Proventriculitis, gizzard erosions and respiratory disease.

Adeno Virus

They are divided into three groups.

- Group I (or) conventional adenoviruses, causes IBH and quail bronchitis.

- Group II includes haemorrhagic enteritis virus of turkeys and marble spleen disease of pheasants
- Group III viruses are found in ducks and cause egg drop syndrome in chickens.

Based on genetic sequencing and cross – neutralization testing, Group I AAV is further divided into 5 species that include 12 serotypes. Nearly every serotype has been reported to cause IBH; however it is most commonly associated with serotype 8a in North America and serotypes 4 and 8a in other global regions. Different isolates of the same species and serogroups will vary in pathogenicity.

Transmission

Both vertical and horizontal transmission plays a role in the spread of IBH. Mostly outbreaks are initiated by transmission of the virus through the embryonated egg and hens exposed during production will typically shed virus to their progeny for 3 to 6 weeks until development of immunity occurs. Horizontal transmission occurs primarily in contact with infected faeces. This seems to occur quite frequently and most cases are not diagnosed because they do not become a clinical problem. Commercial hatching of eggs may be a mechanism of spread of endemic AAV from one area to another. There is an evidence that adenovirus infections can become latent and that period of stress, such as the onset of egg production, will reactivate viral shedding.

Clinical Signs

IBH typically occurs in meat-type chickens under 6 weeks of age but can occur as early as 6 days and as late as 20 weeks. In outbreaks, there is a sudden onset of mortality that usually ranges from 2 to 10% but in some cases, it may be over by 40%. Mortality will vary depending on the pathogenicity of the virus, susceptibility of the chicks (level of maternal immunity) and secondary infections with other Infectious agents. Mortality generally peaks within 3 to 4 days and ceases within one week, although in some cases it may linger for several weeks. Morbidity is low and so the sick birds that are not dead will recover. Affected birds appear depressed with ruffled feathers.

Lesions

The primary lesion is an enlarged, pale and friable liver, haemorrhages may be present in the liver and muscle, in some cases, a straw colored fluid is present in the sac surrounding the heart. In HP, the lesions are similar except the incidence of fluid in the heart sac is greater, Mortality is higher and the age of onset tends to be later (three to five weeks).

Diagnosis

It is critical to isolate the virus from IBH positive samples so that serotyping and molecular analysis can be performed. The most common serologic test is the immunodiffusion test that detects the group I antigen. This test is not sensitive so it is possible to miss positive birds, and does not differentiate by serotype. A group I ELISA is more sensitive but also will not differentiate by serotype. ELISA can be developed for individual serotypes but will not detect the presence of immunity to other serotypes. The serum neutralization test has been used to detect serotype specific antibody but is labour intensive and expensive because 12 serotypes must be included. Interpretation of serologic tests are difficult because AAV is found both in healthy and diseased birds.

Control

- As most cases of IBH are the result of vertical transmission, ensuring that breeder flocks have seroconvert prior to the onset of egg production can prevent the disease.
- Live vaccine and inactivated vaccines have been highly successful for controlling IBH by preventing vertical transmission and inducing maternal immunity.
- Limiting the number of breeding stock suppliers and not mixing stock from different suppliers will help to prevent outbreak of IBH.
- Strict bio-security measures can prevent the disease.

❒❒❒

Chapter-21

Hydropericardial Hepatitis

This disease is also known as Angora disease and Leechi disease. It is one of the emerging diseases affecting the broiler. Disease is quite prevalent in all parts of the world. Adeno virus 1 and many factors are responsible for the disease. The disease appears as major important disease in future. Preventive vaccination at right time is necessary to control the disease. Vaccine must be one type and contains all the inducing factors in right proportion.

During the year 1967, this disease was first recorded in Pakistan at a place of Angora. Since then, this disease is expressed as Angora disease. Afterwards this disease is quite prevalent in all broiler farms in Pakistan. The disease easily spread due to many predisposing factors. High density poultry farm is one of the major predisposing factor. Lack of preventive measures in poultry farm is a basic factor for out break. Existence of other infections like Ranikhet disease, Infectious bursal disease, Avian influenza and other viral diseases reduces the disease resistance capacity of the birds and make the birds easily affected by the infection.

Course of disease is acute in 3 to 5 weeks old broiler. Pericardial sac of the dead birds contain fluids. Mortality rate is much above 50%. This disease is now prevalent all over the world.

Cause

This disease is caused by Adenovirus type 1.

Incidence

Earlier, the disease was prevalent in Pakistan. Due to certain basic problem in the poultry farming, the disease appeared in Pakistan. But now, the disease spreads throughout the world. In the year 1993, this disease was spotted in India near Punjab. Just after 3 years of first appearance, the disease is seen throughout the world. In India, the disease is restricted in Delhi, Pune and Hyderabad during the year 1994 in broiler farm. Affected bird's heart resembles as leechi fruit, hence this disease is termed as 'leechi disease'.

It is difficult to account the attack of the disease, because this disease is quite commonly seen as mixed infections of viral and bacterial origin. This disease causes 50 percent mortality in 3 to 6 weeks birds.

Transmission

This disease spreads both by direct and indirect means. Transmission is direct. Air borne, water and equipment act as mechanical agents in indirect way of transmission. Viral excretion depends upon the antibody level. During the time of reduction antibody level in breeder/layer flock, the virus starts excreting. As the presence of maternal antibody remains for three weeks, the viruses are not excreted. Afterwards the virus starts spreading.

Infectious organisms grow in intestines and multiply in huge number. Droppings of the affected birds contain infectious organisms. Such droppings get attached to the footwear, clothes and equipments easily. Through the movements of these agents, the organism reaches other healthy birds easily. This disease have the capability to transmit through the eggs. Eggs used for vaccine production must be free from infectious organisms. In any case, any eggs contain organism if it is used for vaccine production, results in transmission through vaccine.

Environment is another factor for disease production. Unhygienic environment and lack of biosecurity measures cause the disease. It is possible to locate the disease by symptoms. Increased feed consumption during the initial stages of disease is also a factor. All designed house, nutritional deficiency, uneven growth in grown up birds, mixing of different breeds and raising different age groups and high antibiotic content are some of the predisposing causes for the disease.

Intensity

This is quite commonly seen in the broiler birds aged 3 to 5 weeks. Course of disease remains for 3 to 5 days. Mortality rate is around 10% in each day. Chicks of the layer birds are not affected much.

Symptoms

Different signs are observed. This disease is seen alone and in concurrent infections with other diseases decide the symptoms. Generally, dullness and huddling are seen. Dead birds did not have rigor mortis. Comb and wattles are pale. In the first 3-5 days, the mortality rate is high. In next 3-5 days, shows different mortality rate. In the last 3-5 days, there is a decrease in mortality rate.

Dead birds shows edematous abdomen. Internal organs like liver and kidneys exhibit swelling and broken condition.

Diagnosis

Symptoms and lesions are used for diagnosis, isolation and identification of the causative organism is another important procedure.

Control

Hygienic environment is one of the critical factors. Preventive vaccination is advisable. Prevalence of other bacterial and viral infection should be controlled. Incidence of immunosuppressive diseases like Infectious bursal disease and mycotoxins should be avoided in the flocks. All in all out procedure is better in disease prone areas. Breeder flock should be free from the diseases. Biosecurity measures are quite important in control of this disease.

❑❑❑

Chapter-22

Salmonellosis

Salmonellosis is one of the common diseases in all places. Unhygienic farm condition is the major predisposing factor for cause of this disease. Parent stock transmits this disease through eggs to the chicks. Stress and higher environmental temperature also induces the outbreak. Salmonellosis is a complex disease group caused by many organisms. Symptoms, lesions and outbreak nature are different in different type of diseases. Twenty two types of salmonella species are isolated and identified as a causative factor for Salmonella type of diseases.

Some of the diseases of salmonella are transmissible from man to birds. In the Salmonellosis, transmission of diseases to human beings is an important criteria rather than drop in production in the farm.

Control of Salmonellosis

Infectious organism needs to be identified by blood examination and laboratory investigation. Hatchery hygiene is a critical factor for these types of diseases which are easily transmitted through eggs. In the hatchery, hatchery room, broken eggs and chick hatching tray are to be examined for contamination. Once in two weeks, hatched chicks, faecal matter, different places in hatchery, debilitated chicks are to be subjected to detailed investigation so as to eliminate these types of diseases.

In the brooding places, chick boxes, dead chicks, chicks die within a week, litter, dust, litter in egg nest and inner parts of feeders should be examined thoroughly for the presence of these types of organisms. Even in the slaughter houses all the condemned parts and different places in the slaughter house needs to be investigated. Egg storage place and egg sorting place are to be kept in hygienic manner. It is also equally to avoid rat nuisance in the farm.

Location of Organisms

Breeding flock if contains this type of disease is a major factor for vertical transmission from parent to chicks and horizontal transmission between the birds. Salmonellosis is a major disturbing factor for farm hygiene and public health point of view. Organism present in the faecal matter adheres to the egg shell and transmits the diseases. Day old chicks and live birds easily transmit the disease.

Labourers

Presence of infection in the human beings continuously eliminates the organisms through discharges. Farm labourers also act as an intermediate agency in transmission of disease from infected birds to other healthy birds by various mechanical carriers like clothes, chapels, equipment etc., It is strictly advisable not to allow outside personnel.

Rats, Birds and Flies

Rats not only cause wastage of feed but also responsible for transmission of diseases. White rats are quite common in transmission of diseases like *Salmonella enteritis*. Rats easily transmit diseases from one bird to another.

Flying birds carries infection from one farm to another farm through farm disposals. In addition, flies also transmit the diseases.

Drinking Water and Feed

All possible contamination should be avoided in water storage places. Animal protein sources as feed ingredients are incorporated at 10% level. Animal protein sources like fish meal, meat meal, feather meal, if decomposed usually contains salmonella organism. Sometimes vegetable protein sources like soybean and sunflower oil cake contains salmonella organisms. Possible destruction of microorganisms in the pellet feed is seen.

Control Measures

All possible measures to eliminate breeding flock from having the diseases. Presence of disease in breeding flock easily contaminates the hatchery premises. Litter materials in the breeding flock should be dried in condition. Racking of litter is quite important. Collection of clean eggs without faecal contamination is necessary. Broken eggs and dirty egg should not be used for storage. Minor contamination of eggs can be cleaned properly. Use of disinfectant is advisable in egg sanitation process. Glutaraldehyde at a concentration of 0.5% in the temperature of 40°C for the duration of 3 minutes is necessary in egg cleaning process. Use of formaldehyde at a concentration of 600 milliliter for one cubic millimeter at a temperature of 20°C for the duration of 20 minutes is advisable.

□□□

Chapter-23

Pullorum Disease

Synonym

Bacillary white diarrhoea

Introduction

Salmonella organisms are present in ovary, liver, heart, testes and in many places. This disease was the first identified salmonella type of disease in birds. It also affects ducks, turkeys, pigeon, guinea fowl and sparrow. Outbreak of the disease first occurred in 1989.

Cause

Causative organism is *Salmonella pullorum*. It is rod shaped gram negative and non- motile bacteria with rounded edges.

Transmission

1. Airborne transmission is possible by way of feather dust, contaminated water and flies.
2. Transovarian transmission from parent stock birds to the chicks.

Signs

Important signs are chicks found dead immediately after hatching. Incidence of disease is higher in three weeks old birds. Affected birds are dull, huddled together, closed eyes, reduced feed consumption, pot belly and white bacillary diarrhoea, ruffled feathers and drooping of wings are common signs. Incidence of mortality is 30 to 90 percent in affected flock. Effect of disease is seen for three days. In grown up birds, there is enlargement of hock joints. Feathers are grown poorly and lusterless. Much signs are not seen in layer birds. Infected birds act as source of infection in the farm. Drop in production, infertility, hatchability are other important symptoms. If lungs are involved there is sign of respiratory tract affections.

Lesion

Congestion of liver and lungs are seen in chicks. Affected birds have enlarged belly. Abdomen muscles are inflamed with unabsorbed egg yolk. Depending upon the incidence of disease, the intensity of lesion varies. In per acute type, there is a severe septicemia without significant signs. Affected dead birds shows inflammation of caecal tonsil, lungs, liver, heart muscle and feed grinding place.

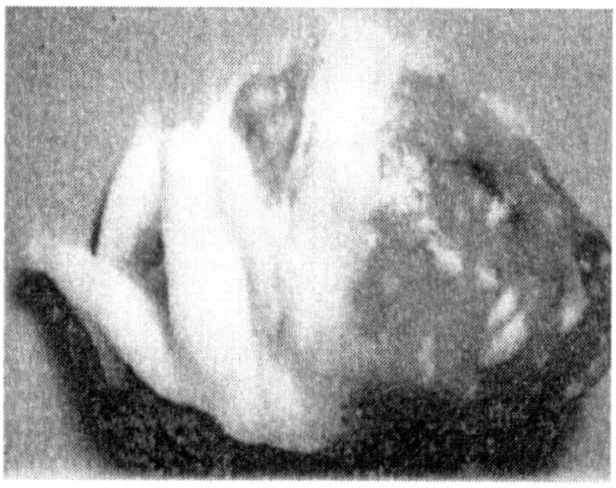

Multiple grey nodules in the heart

In the grown up birds, the ovary are affected. Individual ovarian follicles are misshapen, discolored, distorted truncated and of varying sizes .It appears like bunch of sunken grapes. Egg yolk shows intensity with blood stain. Accumulation of water in pericardium of the heart occurs. In the cock, pericarditis and myocarditis are seen.

Diagnosis

Feed particles present in the intestine are need to be cultured. By antigen and antibody reaction, it is possible to identify the presence of organism in the blood. Antibody titre for this disease is seems to reach titre at 16th week. Plate agglutination test is the easiest way to identify the presence of antibody in the blood. Mixing of antigen with antibody in micro slide and should be examined after 30 seconds. Positive cases exhibit macro agglutination of antigen and antibody which are easily seen as clumps. If the agglutination happens within two minutes, then it is a suspicious. Positive reaction after a period of two minutes should be neglected. It is advisable to repeat the test after a period of one month. 0.02 ml of blood should be mixed with 0.04 ml of antigen in micro slide for conducting the plate agglutination test.

Treatment

Treatment measures vary with purpose for which the birds are used. It is useless to treat the breeding flocks. Even though it is possible to control the disease, it is difficult to avoid the elimination of this disease in breeding flocks. Hence it is not advisable to treat the birds. In this case replace entire breeding flock after thorough cleaning and sanitation measures. Birds used for egg production can be treated with furazolidone for the period of ten days to avoid the effect of the disease.

Preventive Measures

Affected birds easily transmit through the presence of organisms in the faecal matter. This faecal matter contaminates the feed and water, thereby transmits the disease. Salmonella organisms have the capability to be alive in faecal matter for a period of one month. Transmission of disease is possible through egg. By controlling the possible ways of transmission of disease, it is possible to control.

❑❑❑

Chapter-24

Fowl Typhoid

Introduction

Fowl typhoid is caused by the bacteria *Salmonella gallinarum*. It is a gram negative rod shaped and non motile organism. Length of these organisms is 1 to 2 micro millimeters, breadth is 1.5 micro millimeter. Organisms can be alive outside for many days. Temperature of 60°C for the duration of 10 minutes destroys the organisms altogether. The organism can alive for 20 days in water. Phenol at a concentration of 1:100, potassium permanganate at a concentration of 1% for the duration of 3 minutes, and 2% formalin for the period of one minute eliminate the organisms. In the litter without birds, organisms survive for a period of 30 weeks. Incubation period for onset of disease is 4 to 6 days.

Signs

Dull, depressed, increased thirst, drop in appetite, paleness of wattles, comb and head are present. Droppings are yellow tinged. Difficulty in respiration, and increased body temperature are observed. Droppings are adherent to the tail region. Mortality rate is 50%.

In the layers, there is drop in production, dead in shell, embryo mortality, debilitated chicks and dead chicks are primary symptoms. After the onset of disease, birds suffer for 5 to 10 days and end with mortality.

In the turkeys, increase thirst, drop in appetite, dull and depressed nature are noticed. Body temperature reaches 45°C during the course of disease. Just before mortality, the body temperature returns to normal.

Lesions

In acute type, the dead birds exhibit congestion of blood vessels and yellow colour due to jaundice. Enlargement of liver with dark red or black colour are quite evident. Inflammations of intestine and bile mixed intestine contents are seen. Bone marrows are dark in colour.

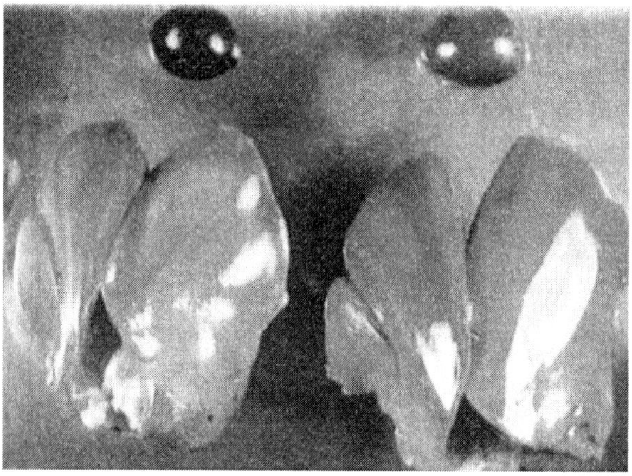

Bronze coloured liver and spleen in contrast to the normal ones at left

Birds died after an affection of long duration. Symptoms such as drop in weight, anemia with necrotic foci in liver, intestine, pancreas and heart, pericarditis, necrosis in lungs and stomach are also seen. Drop in semen production and white necrotic foci in testes are noticed. Ovarian follicles are affected. Egg yolk gets broken and necrosed.

Treatment

Antibacterial drugs are used to safeguard the birds from affection of disease. Furazolidone, an antibacterial agent is necessary to treat the infected flock for the duration of ten days. Enrofloxacin is able to control the disease. Inspite of the treatment, the organisms are excreting continuously. As such, the affected birds are need to be culled and disposed off. Breeding flock should not have any incidence of disease.

In the places of major incidence of disease, vaccination is advisable. GR type of vaccine enhances the immunity. Combination of usage of antibacterials and vaccination are necessary in major affections. Affected birds can be easily identified by taking sample of blood and subjected to antigen antibody reaction.

Control Measures

Even though much preventive measures is adoptable, still certain management procedure is necessary to control the disease.

1. Cleanliness of the farm house is the primary criteria. Use of appropriate disinfectant in the places of disease outbreak is necessary.
2. Chicks should be obtained from the parent stock free which are from the diseases.
3. Entry of disease organisms by farm labourers and vehicles should be avoided.
4. Feed should be free from disease causing agent. Pellet feed using steam destroys the infectious organisms.
5. Rats, bandicoots, flies, mosquito and fleas are involved in transmission of disease.
6. It is necessary to avoid flying birds like crow, sparrow and eagle.
7. Dead birds are need to be disposed properly.
8. Areas of outbreak should be watched carefully.
9. Drinking water should be chlorinated.
10. Due to much importance, the hatchery, feed producer and farm manager all should concentrate in eliminating the disease.

□□□

Chapter-25

Fowl Cholera

Fowl cholera is synonym for pasteurellosis. This disease is caused by *Pasteurella multocida*. This disease occurs in acute and chronic form.

This disease is caused by *Pasteurella multocida*. It is a gram negative non motile, nonsporulated, rod shaped bacterial organism. Many strains cause different type of disease as per acute, chronic and symptomless. 18 types of organisms are identified so far. Organisms are killed by drying and by use of disinfectant and antibiotics.Turkey are easily affected by this disease. Disease is transmitted by air, drinking water and feed. Infection is set up by the entry of organisms through mouth, nose, eye and wounds. Incubation period is 4 to 8 days.

Organism after entry, through epithelium of throat and upper respiratory tract mixes with blood. In the blood, the organisms multiply greatly. Release of toxin is a factor for affection of this disease. It affects blood vessels in all parts of body and induces capillary hemorrhage. Simultaneously there is rise in temperature.

In the plans of higher immunity and lower virulence of organisms, these invasion localizes in parts of comb, wattles, internal organs like hock joint, liver, lungs and brain with much less effects.

Cyanotic comb and wattles

Symptoms

In per acute type, birds die in huge number without any clinical signs. In the acute type, depression, drop in feed intake, foul smelling droppings, cyanosis of comb, wattles, mucus discharge from mouth and noses are important clinical signs. Finally birds die. In the chronic type there is enlargement of hock joint, wattle and comb.

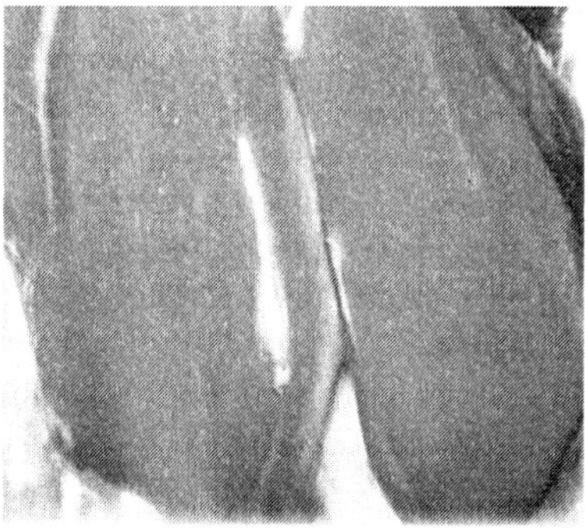

Small areas of necrosis in the liver (corn meal liver)

Lesions

In per acute disease, there is increased capillary petechae in internal organs. In acute type, hemorrhagic lesions are seen in intestine, lungs, pericardium, myocardium, stomach and in all internal organs. Inflammation of lungs is common. Enlargement of liver with necrotic foci are present. This is the pathognomonic lesion of this infection. Impression blood smear taken from affected parts of liver and heart stained by leishman stain shows typical bipolar organisms. Egg yolk broken and ovary parts are affected. In chronic type, swelling of wattles and combs are seen. Affected hock joints contain thick fluid. Inflammation of the lungs and upper respiratory tract are also evident.

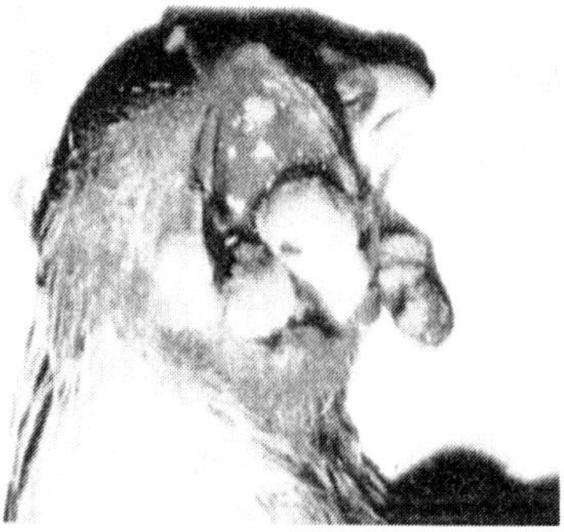

Cheesy exudate in the Infraorbital Sinus.

Diagnosis

1. It is easy to identify the disease by seeing symptoms and lesions.
2. Infectious organisms can be easily isolated and confirmed. It is easy to confirm by seeing typical bipolar organisms in the heart blood smear and impression blood smear of liver stained by leishman stain.
3. This disease is also easily identified by biological tests. Heart blood or liner blood need to be applied on lower part of rabbit abdomen after making scarification in the skin. *Pasteurella multocida* causes death within 24 to 48 hours. Affected rabbits show important sign of hemorrhagic tracheitis

Treatment

It is easy to treat the birds with antibacterials. Ampicillin and sulphanilamide are the drug of choice.

Control measures

1. Carrier birds and animals should not be allowed to nearby poultry farms.
2. Affected birds should be removed immediately.
3. Farm hygiene and equipments are quite important.
4. In continuous outbreak of disease use of preventive vaccination is advisable.

□□□

Chapter-26

E-Coli

Synonyms

Colibacillosis, Hjarres disease, Granuloma, Peritonitis, Pericarditis, Perihepatitis, omphalitis and airsac disease are synonyms of E-coli infection.

Cause

The organism is gram negative, unsporulated and having the size of 2.3 x 0.6 micrometer. Organisms are motile by means of flagella. E-coli organisms are of different types. Three strains O, K and H are available. Strain classification is the criteria for identifying disease.

Transmission

This disease is easily transmissible

1. Droppings containing the infectious organisms contaminate the water and feed. Excess moisture content in the litter is a primary factor. It is necessary to see inside of the waterer both in the cage system and deep litter system of rearing. Sliminess in the waterer is the indication of presence of infectious organisms. Inclusion of

decomposed fish meal having infectious organisms also causes outbreak.

2. Flying birds and rats spread the disease
3. Parent stock transmits the disease to the chicks through egg. Infection of reproductive tract also contaminates egg shell and thereby acts as mechanical transmission.
4. Water is a critical factor in E.coli disease outbreaks. At any point, like collection point, storage point and distribution point have to be kept hygienically.
5. It is also necessary to have hygienic surroundings, Farm hygiene cleanliness and disposal of litter, water and dead birds are carried out in proper way.

Symptoms

This disease affects birds of all age. However depending upon age and intensity of disease outbreak, symptoms vary.

Infection induces changes in egg. Affected egg yolk turns either watery or sometimes thickened with dark yellow colour. Chicks hatched from this disease show unhealed naval cord. Embryo from affected eggs dies before hatching. Depending upon the virulence of the organisms, affection and mortality rate vary. Sometimes there is drop in weight gain and production.

The disease also causes respiratory type of affections. Usually E.coli infection combines with other Ranikhet disease, Infectious coryza and Mycoplasma infection cause greater losses. It is because of viral disease affects mucous membrane of respiratory tract, thereby enables the E-coli organism to enter easily and causes the infection. In such stages respiratory signs are also seen.

Enteric type of infection causes the ovarian affection leads to a drop in production. Involvement of liver results in a drop in weight gain and debilitated condition.

Lesions

Lesions are seen in the affected internal organs. In the air sac involvement, there is swelling and thick mucous discharge in the air sacs. White blood cells get aggregated in these places. Blood accumulates in the lungs. Inflammation of pericarditis is an important pathognomonic lesions. In this condition, the pericardial layer gets thickened, dull coloured and adhered to the myocardium.

Ovarian involvement shows inflammation. Reproductive tract contains thick and foul smelling fluid, such affected flock dies immediately. Egg production also stops. The egg is also having infection. It also leads to peritonitis. Intestines get attached to the peritoneum. Egg yolk is discolored. Affected birds die immediately

In the acute type, liver is affected. Petechial haemorrhage in the liver and perihepatitis are common lesions. Enlarged gall bladder and blood accumulation in the muscles are other lesions. Perihepatitis is the condition, wherein the external mucous membrane in the liver are inflammed and purulent in nature. This is another pathognomonic lesion of E-coli infection. Affection in eyes leads to blindness.

In chronic type, there are nodules in the external part of the intestine. This change depends on birds immunity, type of organism and intensity of the disease.

Diagnosis

Organisms are easily isolated and identified for confirmation of the disease.

Treatment

All broad-spectrum antibiotics respond to the infection.

Control

It is important to identify and locate the path of entry of the E-coli infection into the farm. Control measures depend upon mostly on this factor. Drinking water is one of the most possible ways of the E.coli infection. As such, water hygiene is a critical factor. Quality of the feed is another criteria to avoid infections. Hatchery sanitation invites due attention in preventing the E.coli infection. Diseases are easily controlled by adopting the necessary farm hygienic measures.

□□□

Chapter-27

Infectious Coryza

Haemophilus gallinarum is the causative organism for infectious coryza. It affects upper respiratory tract and contagious in nature.

Cause

Haemophilus gallinarum is a gram negative, non-motile, nonsporulated bacterial organism. Length is 1-3 μm and breadth 0.4 - 0.8 μm. This organism can not live outside for more than 48 hours. In water, viability of the organism is just 4 hours. Outer protective cover resists from the action of disinfectant. It affects the farm production and cause loss. Generally this disease affects the layers. Infected birds transmit the disease through water and air to the other healthy birds.

Pathogenicity

This infectious organism attaches to the cilia in the upper respiratory tract. Outer cell wall antigen is responsible for inducing the disease attack. During the course of growth of infectious organisms, toxin is produced. This toxin is responsible for causing pathogenic changes in the birds. This disease does not affect the chicks which are less than three days. Incubation period is between 1-3 days. Disease duration vary with virulence of the organisms and quantity of infectious organisms. This disease in combination with other diseases like mycoplasma, Infectious bronchitis and Infectious laryngeotrachitis cause much damage. In grown up birds, the duration to cause the disease is less but period of attack is of long duration. Intensity of attack is sometimes to the extent of 100% mortality.

Signs

Typical signs are visible from 7 to 10 days. In mixed infection, period of affection is from one to three weeks. The disease spreads rapidly through the litter. The possibility of attack and mortality are higher side in this condition. Important signs are continuous mucopurulent discharge from nose. In the acute phase, there is swelling of the throat region with inflamed closed eyes. Egg production reduces for 20%. If affected for long period, the disease changes to chronic respiratory disease. Specific rale sounds during the inspiration and expiration are quite evident. In cocks, the intensity of signs are high. In winter season, the birds suffer much higher. The disease is seen from 4 to 8 weeks period.

Lesions

Lesions are evident in the upper respiratory tract. Sinuses in the upper respiratory tract show severe inflammation. In complex problematic cases, the disease expresses lesions and inflammation in lungs and air sacs.

Diagnosis

This disease needs to be differentiated from other respiratory diseases by observing specific signs and lesions. By blood examination, using antigen and antibody reaction the infection can be confirmed.

Treatment

Antibodies and antibacterial drugs at a correct dosage are able to control the disease. Erythromycin and oxytetracycline are the drugs of choice. However antibiotic sensitivity test is necessary to control chronic cases.

Control

In the cases of outbreak, best solutions are to depopulate the infected flock and keep the farm empty for sanitations. By this way, it is possible to eliminate the organisms. Farm house and equipment require to be kept clean using disinfectant. Preventive medication is advisable to control the disease outbreak. Killed vaccine is available which is to be given twice at 4 weeks interval.

This disease is problematic only in ill ventilated farm houses. In the well ventilated areas, incidences of disease of respiratory origin are rare. The cleanliness and farm hygiene is a basic factor to control the disease outbreak in any farm. Dust, cobwebs and daily wastes should be disposed off properly. Surrounding and outside environment are in good hygienic condition.

□□□

Chapter-28

Avian Tuberculosis

Avian tuberculosis is caused by a bacterial organism *Mycobacterium avium*. Primarily the disease appears in the birds after long time production. Unhygienic surrounding is a predisposing cause for the disease. Attack and affection of the disease is of long duration. Organisms are capable of transmitting from birds to human beings.

Infected birds should be culled immediately from the farm. Affected birds are usually in debilitated condition. There is a drop in production. This disease is of zoonotic importance. Affected birds can not be used for food consumption. As the birds are kept in live for long time, the possibility of incidence of this disease is in higher side. The bacterium is alive in soil and other places for long time. It is very difficult to eliminate the organism from the infected place.

Even though this disease is seen in all age groups, incidence is higher in aged birds. In the young birds, only small numbers are succumbed to this disease. In the aged birds, the lesions are specific and intensified. Meat received from affected birds should not be used for human consumption. Sometimes this disease affects the reproductive tract. Egg laid by tuberculosis attacked layer birds contains the organisms. However, the infectious organisms present in the egg do not cause the disease in chicks.

This disease is having long time signs. Attack may be for more time and months. Drop in bird's weight and feed consumption rate is also observed.

Cause

Mycobacterium avium is a causative organism. Organisms are lengthy. Organisms grow in the temperature of 25 to 45°C. Colonies of these organisms are bigger in size. Organisms are non-motile and acid fast in character. There are three types of organisms. Organism seen in human beings does not cause the disease in birds.

Incidence in Birds

Chicken, ducks, goose, peacock, pigeon, turkey and other flying birds are affected. Birds in the zoos have shown the disease. Disease seen in parrot is of human or bovine origin.

All the mammals are easily attacked by this disease. Intensity of disease varies with breeds. Horse, monkey and cats are not attacked. In the pigs, second type of disease is more.

Transmission

Infected birds continuously excrete the organism from the intestine through droppings. Organisms present in the faeces are highly infectious. In the cases of lungs involvement, the organisms are continuously excreted through air. Organisms present in litter and soil infects the livestock. Eggs laid from infected birds contain the organisms. Eggs, as a food processed by six minutes boiling or by 2 minute frying destroy the organisms.

Organisms present in litter get easily transmitted by movements of human beings, feeder and gunny bags. Tuberculosis in pigeon and flying birds easily induces infection in chicken.

Signs

Signs are clearly visible in some birds. Sometimes symptoms are different in many birds.

Affected birds are usually dull, weightless, and isolated from other birds, Even though the feed intake is normal, there is a drop in weight gain. Drop in weight gain causes fat mobilization and subsequent shrinkage of muscles. At last, affected birds are debilitated and weight loss is noticed.

Wattles and comb are pale and cyanosed. Ruffled feathers and dropping of the wings are noticed. Dull debilitated and tiresome look are characteristic signs.

Lesions

Lesions are exhibited in internal organs like liver, spleen, intestine and middle parts of bone marrow. Lesions are seen in all parts of the body. Ovary, heart, testis and skin sometimes show lesions.

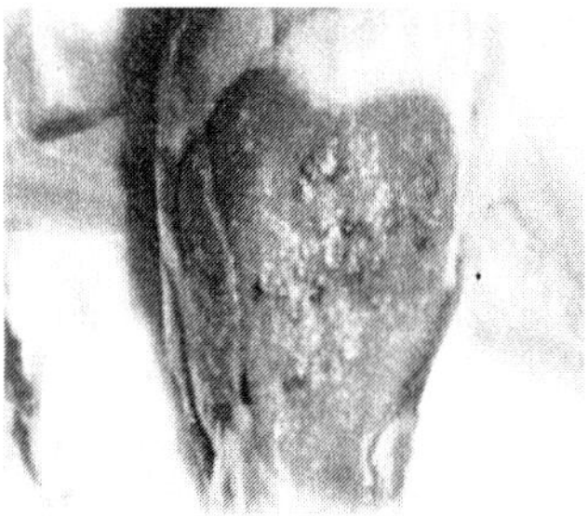

Greyish white nodules in the liver

Affected internal organs like, liver and spleen are grey or greyish white in colour. Tuberculosis nodules vary in size from small size to big size. Enlargement of liver and spleen leads to haemorrhages.

Liver and Spleen

Enlargement with nodules are quite frequently seen. Normal consistency is lost and spleen gets harden. Nodules and pustules present in lurgs are the common lesions in tuberculosis.

Bone Marrow

As that of liver, bone marrow is also affected. Initially there are small size pustules leading to complete destruction of bone marrow.

Diagnosis

a) Specific signs are important to identify the disease. Tuberculin test is very specific to spot out the presence of disease. Tuberculin antigen is injected in wattle and comb region. This reaction is observed for 48 to 72 hours. Positive reactor shows specific reaction in the way of inflammation.

b) Impression smear taken from affected organ should be examined for the presence of acid fast organisms by acid fast staining technique.

c) Affected parts can be injected into the rabbits and observe for onset of disease.

d) Histopathological examination of the affected parts for tuberculosis lesions at cellular level.

Prevention

1. Tuberculosis organisms are get killed by exposure to the sunlight for the period of 3 to 10 days. These bacteria are destroyed by exposure to 2-3% phenol and 1% sodium urea phosphate as disinfectant agents.
2. Positive reactors identified by tuberculin test should be removed immediately from farm premises. In the infected premises, the birds should not be grown after a short interval.
3. Preventive vaccination if necessary can also be used.

□□□

Chapter-29

Chronic Respiratory Disease

This is one of the diseases affecting the respiratory tract. Air sacs are the major parts of attack. Disease is of prolonged course but set slowly. Sinuses of the head are mainly affected. Turkeys are the most susceptible species. The disease transmits from the parent stock to the chicks through eggs. It is difficult to treat the affected flock, owing to the prohibitive cost of the medicines. Exorbitant cost of the medicines to treat the affected birds incur heavy expenditures. Against this, necessary preventive steps are more appropriate.

This disease was identified during the year 1905. Attack of the disease is seen in turkeys during 1938. There are two types of organism causing this infection. *Mycoplasma gallisepticum* and *Mycoplasma synoviae* are two species affect primarily the chicken and turkeys. *Mycoplama meleagridis* and *Mycoplasma jowae* affect turkeys.

Infection in the chickens and turkeys are chronic in nature extending from few weeks to two years. Excreted organisms from the body are alive for few days. Contrary to this, organisms secreted through secretions of the body alive for much longer time. Use of disinfectant destroys the pathogenic organisms. Chicken and turkey are generally affected. Sometimes cage rearing birds and flying birds exhibit this disease. Ducks are refractory to the infection. Disease transmission easily occurs through egg from the infected breeder flock. Infected birds also transmit the infection directly between the birds. In the affected birds secretions of the male & female reproductive tract like seminal secretions and uterine secretions positively harbors the infectious organisms. This disease spread easily through air and also by intermediate agents.

Symptoms

Disease affects the health of the birds. Concurrent infection of other diseases causes much damage. Existence of this disease alone does not manifest much symptoms and mortality. Presence of other infections like Ranikhet disease, Infectious bronchitis, Infectious Laryngeotrachitis, E-coli infections and many other diseases provoke huge damage. Nutritional deficiency disease, presence of ammonia, unhygienic environment, stress and presence of moisture in the surroundings are basic predisposing factors for onset of the disease.

Affection of the respiratory tract exhibits respiratory symptoms. Mouth breathings are observed, depending upon the awakeness of the disease. Eyes are swollen and severe swelling of the eyes partially closes the eyes. Continuous eye secretions and nasal discharge give an ugly appearance. Air sacs in the lungs are generally affected.

Drop in egg production is seen in grown up birds. In the broiler, feed efficiency, poor growth, unthriftiness and drop in weight gain are common signs. Affection of the joints leads to paralysis.

Lesions

As the respiratory tract is affected, lesions are predominant in these regions. Nasal tract, eyes, sinuses around the nose, lungs, bronchi, trachea and air sacs are the sites of lesions. Profuse Nasal discharge is noticed. Sinuses around the eyes and nose have inflammation and swelling. Oozing of the secretions in the affected parts of the respiratory tract is seen. More importantly the air sacs are very much affected. Purulent inflammation is seen adjacent to the air sacs region. Swelling in the airsac is also seen. Involvement of all these organs necessarily disturbs the respiration. Air sacs are also get affected. Generally E-coli infections occur along with other infection. E-coli have inflammation of the liver and pericardial sacs.

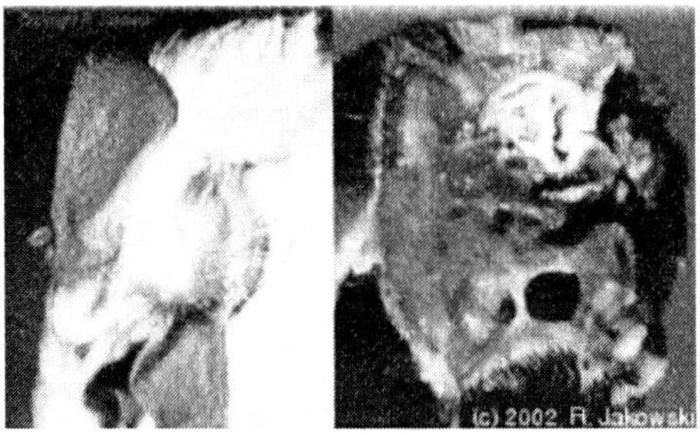

CRD affected fowl

Diagnosis

The disease can be diagnosed by many ways. It is termed as chronic respiratory disease due to prolonged courses, which is the primary point to diagnose the disease. Abdominal respiration and other respiratory sign are important for diagnosis. The diseases do not respond to the ordinary antibiotics or antibacterials. In broilers, drop in weight gain, depressed nature, lacrimation are clear symptoms for diagnosis. Autopsy examination of the dead birds shows specific lesion like purulent inflammation.

Isolation and identification of the causative organisms are the basic laboratory procedure for confirmation of disease.

Serum neutralization test is used for seeing the presence of the antibodies in the infected flock.

Control

It is advisable to adopt all preventive measures rather than treating the affected birds. Elimination of the infectious organisms is the basic step to initiate the control of disease. Owing to the trans-ovarian transmission, preventive measures should be taken at all levels. It is difficult to treat the affected birds. Certain medicines alone are amenable for treating the infected birds. Economically it is not possible to use such costly medicines for treatment. At any point the disease need not be seen in the flock.

Breeder flock should be screened for the infection. It is necessary to adopt plate agglutination test to eliminate the infected birds due to presence of antibody in blood at the age of 6th and 18th week. Before the egg production the flock should be checked thoroughly. Eggs received from the disease free flock must be used only for hatching purpose. If it is present in the hatching eggs, certain procedures are taken up to eliminate the infectious organisms. One of the methods is the dipping of eggs in the medicinal fluid. Hatchery environment must be in hygienic condition. Introduction of medicines into the egg albumen eliminate the organism. Heating of the eggs to the temperature of 46 - 47°C also control the infectious organisms. At any account, the infectious organism should be eliminated from parent stock and chicks.

Immediately after hatching the chicks should be given medicated drinking water alone for the first three days. Owing to the lesser quantity of water consumption, the medicinal cost also low. Depopulation of the farm for certain period is a correct approach to control the disease, as prohibitive cost of medicines does not allow the treatment. During the period of depopulation, thorough cleaning is necessary. Disinfectant like formalin is better due to its action in air. Hygienic environment and good surroundings are critical factors for the control of diseases.

❑❑❑

Chapter-30

Protozoan Infection Coccidiosis

Coccidiosis is one of the important protozoal diseases in poultry. This organism multiplies in the intestines and thereby causes injury. It leads to a drop in feed consumption and digestibility, subsequently leads to excretion of water and blood. It further complicates with secondary bacterial infections and loss of profit. Coccidiosis affects the younger birds mostly. This is actually due to the establishment of immunity in birds. Cross immunity is not seen in coccidiosis. Shorter incubation, direct life cycle and prolific reproductive tendency nature makes this disease out break as worst. The disease outbreak is due to few oocysts or few million oocysts either through water or in feed. Vaccine for this disease is not used due to vaccinal reactions in the birds as production drops and growth retardation.

Coccidiosis is caused by nine diffèrent species of coccidia. However, major affection is due to *Eimeria tenella* and *Eimeria necatrix*. Idenfication of coccidial species is done by following factors.

1. Intestinal lesions
2. Intensity of pathological lesions

3. Shape, measurement and colour of oocysts
4. Nature of schizonts and merozoites in the intestines.
5. Location of disease producing organism
6. Incubation period
7. Sporulation time
8. Disease resistance.
9. Modern chemical methods can able to identify the organisms exactly.

Eimeria Acervulina

Oocyst is round in shape and the narrow end is softer in nature. Intensity of attack depends on the type and number of organism and immune status of the birds. Retardation of growth rate depends on number of organisms. Moderate and less attack does not affect feed efficiency and weight gain. However, affected birds are anaemic with yellow colour discoloraton. Yellow colour discoloration is due to impairment of absorption of vitamin A in the intestine. Intestinal mucous membrane gets thickened leads to poor feed efficiency. Drop in egg production is also seen in layers.

Lesions

Pathological lesion is clearly visible on the serosa of the intestines. During initial period, intestine gets thinned with whitish materials. These types of changes are observed both across and lengthwise in the intestine resembling a staircase appearance. Sometimes intestines appear pale with watery contents. In lesser intensity of attack, pathological lesions are seen in duodenum, but in severe attack lesions are seen in all parts of intestine.

Eimeria brunetti

Oocyst of *Eimeria brunettti* resembles *Eimeria tenella*. It affects the last part of the intestine ileum. Severely affected birds show lesions in small and large intestines. Confirmation of the disease based on the lesions in the intestine is difficult. Affected birds have lesser degree of reduced weight gain, poor feed efficiency and mortality. Intensity of attack is not so severe in *Eimeria necatrix* and *Eimeria tenella*. 100 to 2,00,000 oocysts induces mortality percentage in the range of 10 to

30%. Recovered birds have poor weight gain capacity. Lesser intensity of attack goes undiagnosed mostly. It results in minor reduction in weight gain and feed efficiency. In the initial stage petechiae is seen in the mucosa of the intestine. Thickening of the intestine results in change of colour. Necrosis of the mucosa of intestine is witnessed in severe attack. Blood clots are seen inside the intestine and mixed with droppings. Swelling of the mucosa of the intestine is also seen.

Eimeria maxima

Middle part of the intestine is affected here. Severe attack results in affection in all parts of the intestine. Oocysts are bigger in size and are easy to identify. Oocysts are yellow in colour. Intestinal contents in middle part of the intestine are yellow in colour, hence oocysts present here are also yellow in colour.

This type of coccidiosis causes moderate attack. Infection induced by 2,00,000 oocysts results in poor weight gain, diarrhoea and mortality. Affected birds are debilitated, anaemic and ruffled feathers. Feed intake is also reduced. Affected birds show poor assimilation capacity, resulting in increased depigmentation of skin. Swelling of the intestine along with watery contents leads to ballooning of the intestine. Lesions are not much observed in this type of coccidiosis.

Eimeria necatrix

This type of coccidiosis is widely seen in all parts of the world. Path ognomonic lesions are clearly seen in this disease. Intestinal contents are mixed with blood. Enlargement of the intestine is seen. Oocysts are seen only in the caecum.

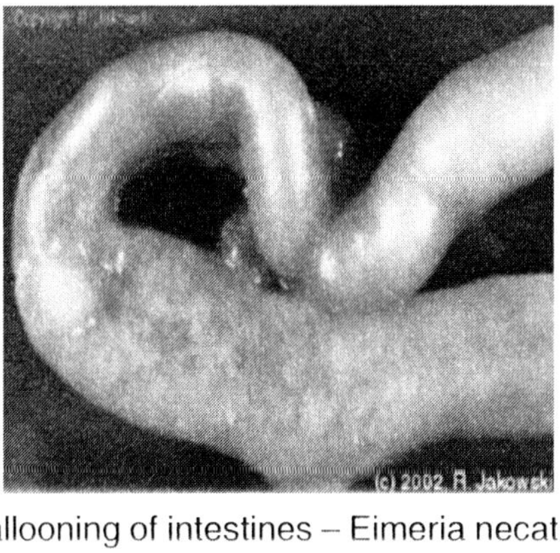

Ballooning of intestines – Eimeria necatrix

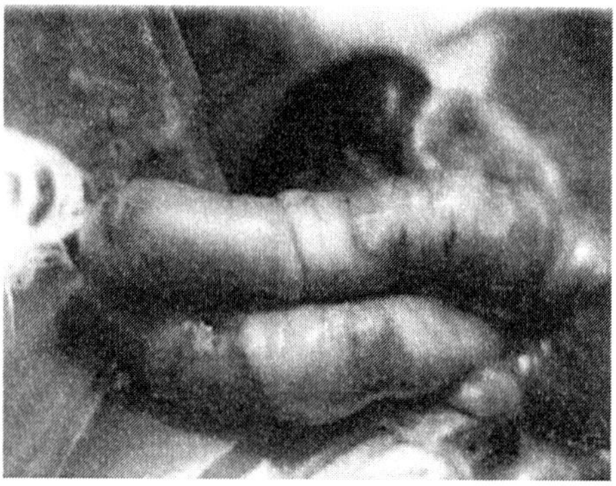

Intestinal distention, mucoid blood filled exudate and white spots noted on the serosa with E. necatrix infection.

Symptoms of the disease are evident after 56 hours of sporulated oocyst's indigestion. Ballooning and thickening of the intestine along with digesta mixed with water, blood and intestinal tissues and can be easily confirmed by identifying the petechiae and white spots on the serosa of the intestine. Affected parts of the mucosa of the intestine on microscopical examination reveals the presence of merozites and schizonts. Schizonts are responsible for haemorrhagic lesion in the intestine. Sometimes in severe attack, swelling of the intestine is observed. Infected bird's dropping is mixed with blood. Coccidial oocysts in the range of 75 to 1,00,000 oocysts induces reduction in body weight gain and mortality. Recovered birds are emaciated and anaemic.

Eimeria tenella

Caecal coccidiosis is very common. Broilers are usually affected. This organism affects the caecum. Affected birds show mortality and severe reduction in weight gain. This type of coccidiosis causes severe losses. One lakh coccidian oocyst induces severe mortality and weight loss. Attack due to 1 to 3,000 numbers of coccidial oocysts results in bloody droppings. Second generation schizont is responsible for pathological changes in the intestine. Affected parts of the intestinal mucosa have intermediate stages of merozoites and schizonts. Hemorrhagic lesion is due to rupture of second generation schizonts in the intestines. Schizonts grow in the deeper parts of the intestine, process of release of merozoites from the schiozonts results in inflammation of the intestine and consequent bleeding. Mortality is seen between 5 and 7 days of attack. Weight loss is maximum on 7th

day of infection. Mortality is due to toxic substances and bleeding. Extract of the affected part of the intestine when injected into the healthy birds is able to create blood clots and mortality. Mortality could be also due to secondary bacterial invaders.

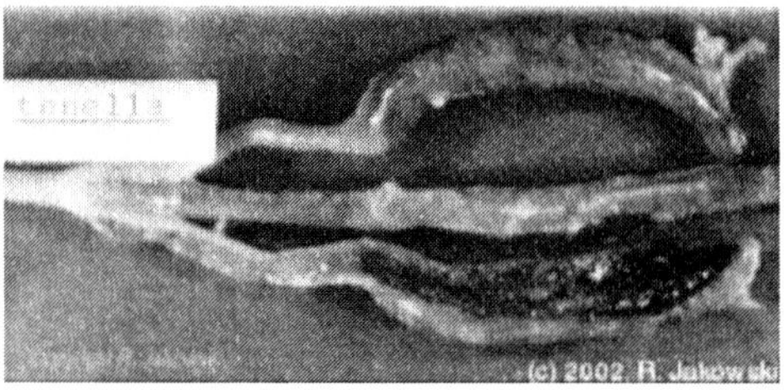

Lesions

Second generation schizonts appears after 4th day of the infection of the sporulated oocysts. These schizonts are responsible for pathological lesions. Blood clots are seen in the caecum. Recovered bird tissues are replaced with new tissues. Bluish colour, strong petechiae is very clearly seen in the serosa of the intestine. Swelling and increased thickening of the caecal wall are also possible.

Transmission

Infection is feasible only with ingestion of viable sporulated oocysts. Infected birds discharge oocysts from 7 days to several weeks. Excreted unsporulated oocysts gets sporulated within 48 hours and become infective. Pecking behaviour of birds favors intake of oocysts. Between farms, transmission of disease occurs through dust, livestock, insects and equipments. Oocysts are unaffected by environmental changes and disinfectants. But viability of oocysts get reduced. Oocysts are viable for many weeks in the soil.

Control and Prevention

As soon as the disease is identified in the farm, immediate attempt should be made to treat the birds with specific drug/medicines. Prevention is better than to treat the birds. It is necessary to adopt suitable preventive measures to arrest the disease outbreak. Drugs used should be checked for its action whether it is for killing or arresting the growth of the organisms. Medicines may have specific action either against one oocysts or on all oocysts. Selection of right drug is important. Each and every drug is having its own specific actions.

Sulphanilamide type of drugs interferes the metabolism of the infectious organisms by interfering with folic acid synthesis. Amprolium type of drugs controls the infection by preventing the absorption of thiamine by the parasite. Clopidol type of drugs act by affecting energy synthesis from metabolic process. Polyether ionophore drugs control the infection by interference with osmosis

Coccidiostats act by inferring with life cycle at different stages. Drugs acting at initial stages will not induce immunity. Such drugs can be well used in the broiler farming. However coccidial immunity is necessary in the layers. In that case, it is advisable to use drugs which act in later stage of the life cycle. Drugs, which kills the organism will not favour immunity establishment. Drugs having growth inhibition action can initiate immunity. Coccidiostat type of drugs induces immunity whereas coccidiocidal drugs prevent immunity.

It is also advisable not to use one medicine continuously. Because such continuous usage can lead to development of drug resistance. In such cases, some drugs may be totally ineffective. However, it is important to see establishment of drug resistance. Now, coccidial vaccines are available, still it is uneconomical to use on wider areas.

Critical moisture content in litter should not increase beyond 20%. Moisture is important for sporulation of ooysts. Hence litter management is altogether important. Coccidiosis outbreak indicates the resistance of unhygienic conditions in the farm.

□□□

Chapter-31

Ectoparasites

Ectoparasites come under the classification of Arthropods in animal kingdom. These parasites live in the skin and body of the living organisms. It also multiplies in walls, faecal droppings and dead carcasses. In chicken different types of ectoparasites are seen.

Ectoparasites live in the dead cells and organs. Most of the parasites live on the skin. Inhabitant in skin region maintains body temperature. Some ectoparasites are nocturnal in habit.

Body of the ectoparasites are divided into three parts as head, neck and abdominal region. In the head region, there are two antennas and three pairs of legs are seen in the body. Some ectoparasites contain wings.

During the time of infestation of ectoparasite birds starts picking and scratching to avoid the irritation caused by the ectoparasites. Sudden drop in production and higher feed expenditure could be the reason for ectoparasites. Good light and vision are necessary to identify the presence of ectoparasites in strain. At least 50% of the birds are need to be screened for seeing the ectoparasitic problems. All birds are need to be examined. External parts like head, neck, wing and legs are to be checked.

Blood parasites are difficult to identify in the birds. Poultry housing, walls, crevices in building and poultry dropping are to be examined for blood sucking parasites. Needles can be used to spot the parasites in the crevices of woods. Uncared dust present in the farm and equipments for some periods can be exposed in the white paper for spotting the ectoparasites by its movement.

Preventive Measures

There are different ways to control the parasites. Affected birds need to be dipped in the medicated water by dusting with acaricides. Bird's skin are well exposed to the medicated water or powder thoroughly including its hidden parts. Eradication of the parasites in the building necessarily requires thorough spraying with medicated water. There are different ways to use the acaricides.

1. *Powder*: According to the area, the required acaricide powder is sprayed to the floor, walls, roofs, feeders and all parts of the farm. Sometimes dust bath is used. In the dusted litter material contains specific quantity of chemicals. During the act of sitting, the skin of the bird comes into contact with the chemicals.

2. *Spray*: Pressurised spray is useful. During the usage of spray, the pressure to be maintained is at a level of 50 pounds per square centimeter. Powerful spray is advisable. In the usage of spray, the acaricides is mixed with water and then it is used.

3. *Mist*: It is possible to control the ectoparasites by this method. This spray works on very high speed, powerful and operated easily. Concentration of the chemicals is 5 -10 times higher than spray type. Acaricide water must be thoroughly well mixed to have good effect.

Lice

Different types of lice are seen in birds. Lice have mandibles. Mandibles are flat and have small sensitive antenna, 40 types of lice are identified so far. All types of lice can be prevented by proper preventive measures. Sometimes the birds may contain more than one type of lice. Menocanthus straminus, Menopan gallinae and Liperus numidae are the common lice seen in birds. Lice are generally seen in skin and feathers. It is necessary to have skilled view to spot the lice. Lice are having a length of 1-6 mm. Lice complete its lifecycle in the birds itself. Eggs of lice are attached to the feathers. Eggs need 4-7 days for hatching. Life cycle of lice is 3 weeks. Single lice can produce 1, 20,000 lices. These lice live for few months only. Mostly lice live by eating the feathers. Lice burrow the feathers and sucks blood. Lice infestation causes drop in weight gain and nutritional deficiency. In birds, drop in egg production, delay in egg laying period and increase in feed cost are some of the ill effects of lice infestation. Lice do not transmit the disease, but irritates the nerve endings thereby creates restlessness in the birds. Lice infestation also affects the health of the birds. Encephalomyelitis virus of horses is isolated from lice. Still there is a report that diseases are transmitted by lice.

Control Measures

Lice infestation is common in winter and rainy season. It is advisable to adopt necessary preventive measures during this season. Birds should be treated at an interval of 10 days. No one chemical can kill or eliminate the egg of the lice. As such the birds need to be treated again lice to newly hatched lice.

Generally ectoparasites exhibits in the feathers. Dropped feathers contain no lice. Farm house should be maintained in hygienic atmosphere. Sprayed chemicals should reach all parts of the farm house. Birds neck, abdomen and tail region are to be checked for entry of medicated water. Caged birds should be checked frequently for lice. Twice a month, the birds should be examined for lice. By implementing the preventive lice infestation even at a minimum level can be controlled.

Fleas

Matured fleas are seen in the birds. Length of the matured fleas is 1.5 to 5mm. Flattened body on sides , sucking mouth parts, small hair like antenna and lengthy legs are some of the characteristic features of the fleas. Mostly black or dark brown in colour. Fleas daily lay many eggs. Hatching of eggs requires warmthness. The Larvae come out after a period of two weeks. Larva lives on droppings and decomposed matter and eats the feather and outer layer of the few matured larval brood to reach the next life of pupae. These pupae live for one or few months. From pupae, flea hatches out and inhabitates in the skin by sucking the blood on their feed. It reproduces in that place. Matured fleas are capable of living for few weeks without feed. Fleas live in the body of the birds for few months. Room temperature, moisture content and environment accounts for the total life period of the fleas. *Echinophaga gallinacea* is otherwise known as Stick tight flea which is the common flea seen in poultry. This type of fleas are numerous in the head region. Mouth parts hold the skin. Matured fleas lay eggs in the litter.

Prevention

Matured fleas should be eliminated by using sound and good acaricides. In the infected farm, replace the old litter with new litter.

Mites

Mites have hard back and abdomen. Mites are having a tendency for drinking blood. Mites live by causing injury in the feathers. There are 25 types of mites. Infested birds show dullness, drop in egg production and many other symptoms. Cnemidocoptes mutans affects the legs and cause scaly leg condition. Adult birds generally show this problem. This type of mite has small legs and striated body. Diameter of adult mites measures about 0.5 millimeter. Mites completes its life cycle in the skin. Infestation between the birds is due to contact. These mites are generally seen in featherless parts like leg and head region and burrows in the leg causing swellings. *Cnemidoscoptes gallinae* is another common mite seen in the feathers.

Prevention

Affected birds should be separated first by dipping the birds in medicated water along with thorough spraying of good acaricides.

Ticks

Ticks measure about 3-4 millimeter in length. Female tick is lengthier to extend about 10mm. Dorsal part of the ticks are not hard. Ticks outer part is tuber like texture with shrunken knob like appearance. Head is under the front side body. Many damages are caused by the ticks. Anaemia and intoxication by the injected toxin occurs. Tick transmits disease like spirochaectosis and piroplasmosis. Argus persicus is the common poultry tick seen in all places. These ticks can cause paralysis in birds due to presence of paralytic toxins present in them.

Prevention

All the places should be free from growing and mature ticks. Crack and crevices should be checked for ticks.

□□□

Chapter-32

Endoparasites

Among parasitic diseases, endoparasites constitute a most important place. Endoparasite grows in the internal organs, thereby affects the production and health of the animal. Most of the endoparasites are present in the intestine, hence termed as intestinal worms. Endoparasites are classified into three categories. They are round worms, tape worms and flat worms. In general intestinal worms are affected by two ways.

1. Intestinal worms grow by taking required nutrients from system, resulting in depletion of nutrients.
2. Enteric worms affect the organelles function, where it grows. Sometime it produces toxic substances to damage the parts concerned.

Due to decreased nutrient contents, growth and production is affected. Further it induces stress, leads to attack of pathogenic organism.

Round Worms

Generally, roundworm affects mainly the organs of the digestive system. Roundworms are of many types. Some requires insects as intermediary stage to complete the life cycle. Higher number of round worm produces major changes. However, commonly seen round worms are few only.

Worm Affecting Respiratory System

Syngamus trachea affects respiratory system of the birds. *Cythostoma bronchilis* affects goose.

Affected birds have mouth breathing with extension of neck. Continuous affection of these worms causes mortality. Transmission is possible either directly or through intermediate host earthworm. Affected farms soil continues to harbour the eggs for many years. Free range birds have this worm commonly. It is necessary to differentiate *syngamus trachea* eggs from capillaria spp.

Enteric Worm

Ascardia, Heterakis and Capilleria constitutes three important worms. It is easy to differentiate these worms by seeing the morphological features.

Capillaria sp worms (hair worms) are the smallest among enteric worms. It is a long thread like worms found in oesophagus, crop, small intestine and caecum. Greater intensity of affection is seen in higher number. This worms transmits the disease either directly or indirectly. There are many types. Depending upon the worms intensity of attack varies in different parts of intestines. Deep litter reared birds continues to have capilaria. Microscopical examination of the liquid contents of affected intestine mucosa can reveal the presence of these worms.

Heterakis gallinarum is seen in the caecal part of the intestine in chicken and Turkey. This type of worms does not cause much problem. However, this worm is responsible for transmission of one of the important protozoan disease Histomoniasis. This worm can be infected by direct transmission.

Ascardia type of large round warm affects chicken. Site of Prediction of this worm is small intestine. Larva of this worm enters the intestinal mucosa and thereby induces problem. Depending upon the number of worm, intensity of attacks will vary. Affected birds have enteritis and sometimes intestinal blocks are seen. Transmission of this worm is direct. Apart from these, trichostrongylus worm also affects chicken and free flying birds.

Tape Worm and Flat Worm

Many types of tape worm and flat worm are commonly seen in chicken. These worms do not cause much damage. As infection occur in chicken, intermediate host snail is necessary. Commonly these worms are seen in free range birds. Arthropods act as intermediate host for tapeworms.

Railletina, Cysticercus, Davaenia proglottina are the common tape worms. Railletina tapeworm requires ant as its intermediate host, whereas snail act as intermediate host for Davaenia worms. Railletina, Cysticercus grow in the digestive and reproductive tract and disturbs the function.

Control

Poor management practice is the predisposing factor for worminal infestation. Broilers are not usually affected, because, they are grown for shorter period. In intensive system of rearing, spread of infestation is possible from grown up birds to others. It is difficult to eradicate ascardia infection. Reuse of litter favours the spread of infestation is possible from growing birds to others. It is difficult to eradicate ascardia infection. Reverse of litter favours the spread of infestation.

Control measure is not possible in wild birds and free range birds. Hence affected birds should be treated first, before taking up the preventive measures. Rational for treatment should desire upon action of drugs as wormicide and wormistat.

It is also necessary to have the right approach of eliminating and killing the intermediate host. Elimination of worm without killing eggs leads to further outbreaks. Right drug and dosage with fixed schedule is essential to contact the disease.

Levamizole or piperazine medicines can be used to treat the roundworm. Thiabendazole and mebendazole type of drugs can be used for treating tapeworm and syngamus trachea.

□□□

Chapter-33

Mycotic Diseases

Brooder Pneumonia

Respiratory affection due to brooder pneumonia is a common feature in low threshold managed farm. Chicks, duckling and turkey poults are affected by this disease. High ammonia levels, cold shock and dust are the predisposing factors.

Birds in high intensity farms are easily susceptible. Lack of resistance, septicemia and nutritional deficiency diseases make the birds prone to this disease.

Etiology

Aspergillus fumigatus is the causative organism. *Aspergillus flavus, Aspergillum sniga,* and *Aspergillus gallus* are also isolated from the affected birds. This fungus grows well in organic matter having more than 22°C temperature. Fungus is 4 to 12 mm diameter.

Transmission

Spores present in the breathing air enters lungs and cause the disease mainly, this disease spreads from hatchery. Infected eggs in setter explodes, leads to floating of large number of spores in the air. Contaminated vents and equipments are the source of infection in setter and hatcher. This infection is directly transmissible to the just hatched chicks. Spores present in the feed and litter also aids in transmission of disease through inhalation of air. Wet litter favours multiplication of spores, subsequently during drying process, the spores gets floated in the air.

Mortality occurs at two times in the course of infections. Initially, inflammation process of lungs leads to mortality in the first three weeks. Later due to impairment of function of lungs, ascites, loss of sight and nervous dysfunction results in another 5% mortality.

Symptoms

Affected birds have distressed breathing. After 3 weeks, nose blocks lead to mouth breathings. Recovered birds appear dull and depressed, weak, Swollen eyes with loss of sight and nervous weakness, grown up birds have lowered respiration efficiency and unable to cope up with production efficiency. Sometimes laryngeal flakes predisposes to mortality.

Lesions

Pathogenicity is due to growth of conidia in the alveoli. Unnecessary tissue growth disturbs functional activity. Creamy white nodular growth of 1-15 mm is seen in the surface and parenchyma of the lungs. Center of nodules have fungal growth with necrosis.

Diagnosis

Early diagnosis is possible by observing symptoms and seeing the pathognomonic lesions in the lungs. Confirmation of the disease is by isolating the organism from the nodular growth seen in the lungs. Differential diagnosls is necessary from E-coli, staphylococcus and salmonella organism. Microscopial examination of the affected tissues can reveal the presence of fungal organism. Morphology and chemical characteristic of the organism are need to be correlated.

Prevention

There is no treatment for this disease. Attempts should be focused to prevent the disease. Spread of spores should be avoided. Caution should be taken to avoid the spread of infection through the egg. All equipments should be thoroughly cleaned. Wet litter problem should not be there. Feed should be free from dust and fungal attack. Feeders and waterers should be cleaned with suitable disinfectants. Chronically illed birds have yellow or black coloured airsacs. This is due to growth of fungus. Fungi grow continuously and affect the surrounding tissues.

Chronically affected birds have damaged tissues leads to disturbances in blood circulation. Further, it is complicated with ascites. larygeal block is observed most of the times in dead birds. Spores enter the blood circulation and affect other internal organs like brain, heart, bone, kidneys and other places. Growth of fungus in brain complicates the infection.

□□□

Chapter-34

Mycotoxicosis

Introduction

Effect of mycotoxin on the health of the human being was first recorded in 1975. In this report, ergot toxin was produced by a fungus *Claviceps purpura* is the causative factor. In 1942, enteritis due to mycotoxin was reported in Russia. In 1960 mycotoxicosis was seen in turkeys.

Toxin produced by fungus is termed as mycotoxin. Disease condition caused by this mycotoxin is called as mycotoxicosis. So far, 23 types of mycotoxins are identified. Each fungus produce individual mycotoxins. *Aspergillus flavus* produces aflatoxin. *Aspergillus ochraceus* is the organism for production of ochratoxin. Like way, there are different types of toxin as rubratoxin, L2 toxin, zearalene, citrinin, oosporin, cyclopiazonic acid and strigmatocystin etc., Presence of mycotoxin in feed lowers the quality.

Mycotoxin affecting Poultry

Mycotoxicosis is the most prevalent problem in poultry. Mycotoxicosis is caused by a toxin of single identity or by complex identity of more than one toxin. Much loss was occurred due to aflatoxicosis. Now incidence of ochratoxin A and cyclopyosonic acids are also reported. As such, it is necessary to test the feed quality for the presence of more than one toxin.

Mycotoxin in Feed

Growth of fungus happens either before or after the harvest or at the time of harvest. It is entirely due to external environmental temperature and moisture content. During the growth of fungus, the toxins are produced at various stages. Mycotoxin is seen in the following feed ingredients.

Oil cake	-	Groundnut oil cake, cotton seeds, coconut oil cake, Sunflower oil cake and soyabean oil cake.
Cereals	-	Maize, sorghum, rice, wheat, barley and peas
Tubers	-	Potato and tapioca

Mycotoxin contaminated feed ingredients in the course of feed mixing enter the poultry feed. Even good quality feed, if not properly stored is prone for growth of fungus and subsequent production of mycotoxin.

Mycotoxin Types

i) *Aflatoxin*: Among the mycotoxins, aflatoxin induces greater damages. Aflatoxin is classified as B_1, B_2, M_1 and M_2. Aflatoxicosis is a common condition after the rainy season in cattle and chicken. This toxin primarily affects the liver.

   This toxin affects metabolic process of carbohydrate. It reduces the potency of the liver enzymes. At cellular level, this toxin interferes with Krebs cycle in mitochondria thereby affects the health of the birds. It also affects the metabolism of fat and protein. Apart from ill health and production drop, it reduces the body resistance and immune levels.

   In the ducks, the level above 0.02 mg per kg of feed causes aflatoxicosis. In chicken, the maximum permissible level of aflatoxin B_1 toxin is 0.2 mg per kg of feed. Ducklings and turkey poults are severely affected than chicken.

   Layer feed containing 100 ppm of aflatoxin B_1 causes 5% mortality and 95% drop in egg production within 48 hours. Feed having 20 ppm of aflatoxin $B_{1,}$ if fed to cocks induces off feed, drop in weight gain and drop in semen production and quality. Level of 1.5 ppm of aflatoxin B_1 in broiler mash causes poor growth from first week to eighth week.

ii) *Ochratoxin* A: Ochratoxin A causes damage next to aflatoxin. It causes much serious damages in western countries. Feed containing ochratoxin above 2 ppm only cause the damages. At lower level of 0.5 ppm, it kills the pathogenic microorganism in the intestine.

iii) *T_2 toxin*: This toxin is produced by Fusarium Spp. This toxin affects growth and digestion of birds. Affected birds shows production drop, skin changes and fall in egg production are observed.

iv) *Zearalene*: This type of toxin is present in maize and sorghum. It affects abdominal region. Blockage of reproductive tract and orchitis are common symptoms.

v) *Citrinin*: This toxin affects kidneys. Increase in water consumption by this toxin leads to watery droppings.

vi) *Vomitoxin*: Regurgitation of feed is seen in this toxin.

vii) *F_2 Toxin*: It causes embryonic mortality.

viii) *Oosporin*: Basically affects kidneys. It also induces paralysis.

viii) *Cyclopiazonic acid*: This toxin is seen in the maize. This toxin reduces egg production.

Factors for Production of Mycotoxin

Generally for the toxin production environmental temperature should be 21°C & relative humidity should be 85% and moisture content in feed between 13 to 15% are required. Problem is more in rainy season. All ventilated places are conducive for growth of fungus and subsequent production of toxin. Pests and beetles in the feed damages the feed ingredient thereby increases the temperature and also moisture content of feed ingredients. In these conditions, fungus favourably grows and produces toxin. Beetles also transfer the spores of fungi from one place to another place. Insufficient attentions in feed storages give a favourable condition for the production of mycotoxin.

Preventive Measures

Inclusion of toxin in the feed or feed ingredients can be exerted by two conditions.

1. By controlling fungal growth.
2. By eliminating the toxin in the feed.

Ways to Control Fungus Growth

- To eliminate fungal growth before harvest, necessary scientific procedures should be adopted. Fungal growth resistant strain should be advocated.
- Use of drugs to control the fungal growth.
- Appropriate and systematic harvest is necessary. But this way, it is rather possible to control moisture content and damage of feed ingredients.
- Right way of storage with good utilization is appropriate. Infestation with flies and beetles should be avoided at all levels.

Means to Eliminate Toxin

Many methods are available to eliminate toxin.

1. By use of necessary chemicals
2. By exposure to sunlight and heatings
3. Microorganisms
4. Biological method

Still the cost effective method to eliminate the entire toxin in the feed is not available right now. Exposure of aflatoxin contaminated feed for six hours of bright sunlight reduces 60% of aflatoxin content.

❒❒❒

Chapter-35

Nutritional Deficiency Diseases

Nutrients are the critical factors in poultry farming. Nutritional deficiencies are caused by inadequate supply of nutrients, indigestible form of nutrients, inactivation of nutrients and non-utilisation of nutrients in feed. Each nutrients account for some physiological function in the body. Vitamin A and K keeps the peripheral part of the cells in healthy condition. Vitamin K involves in the coagulation of the body. Water soluable vitamins are responsible for some of the physiological functions of the body.

Nutrients in the body are classified in many ways. Deficiency, excess level and unbalanced form induces many changes in the body. Nutrients are classified into four main parts.

1. Higher levels of nutrients carbohydrate, protein, fat and water.
2. Important major minerals like calcium, phosphorus, potassium, sodium and chloride.

3. Fat soluble vitamins like A, D, E and K and water soluble vitamins like thiamine, riboflavin, niacin, folic acid, pantothenic acid, pyridoxine and Vitamin B_{12}.
4. Trace minerals like iron, calcium, copper, manganese, zinc etc.

Some of the nutrients are deficient basically in feed. Some of the unwanted materials like antinutrients, toxins and infectious organisms are responsible for inducing deficiency condition. Balanced feed and water is the basic requirement for any living organism. Generally in birds for good production and growth, 36 essential nutrients are required. It is imperative to differentiate from other diseases of poultry. Poor and inadequate growth, poor production, effect on hatchability, dropping of feathers, nervous system affection is caused by some nutritional deficiencies. Each nutritional deficiency characteristically manifests specific symptoms.

Protein Deficiency

Protein is an essential nutrient for production and growth. Protein is building up of aminoacids. Aminoacids are of 20 types. The amino acids like arginine, histidine, isoleucine, lysine, methionine, phenylalanine, threonine, tryptophan and valine are classified as essential amino acids. Amino acids like cystine, tyrosine can be prepared from other essential amino acids. Chicks require amino acids like glycine or serine and proline. Preparation of non-essential amino acids requires protein.

Generally poultry feed is deficient in one or more amino acids. Maize and soybean based chick brooder mash contains lower amount of methionine and lysine. Cereals, groundnut and gingelly oil cake based feed are lower in methionine and lysine content.

Deficiency of amino acids does not manifest specific symptoms. Usually, retarded growth, drop in egg production, reduced weight gain, unthriftiness and poor feather growth are the generalized symptoms seen. Protein deficiency either increases the feed intake or accumulates the fat deposition in the body. Deficiency of amino acids affects physiological functions of the body. Methionine deficiency increases choline and vitamin B_{12} deficiency status. Lysine deficiency causes lack of pigmentation in turkeys. Argenine deficiency bends the tail point and makes it rough in texture. Many amino acid deficiency changes the feather structure and growth.

In the higher protein level during the process of metabolic activity, urea nitrogen is released as uric acid. Like this, urea creates gout condition in chicken.

Carbohydrate Deficiency

Major part of the feed is utilized for metabolic heat. Birds digest sucrose and starch, but unable to digest lactose. Owing to this, milk and milk products cannot be incorporated in the feed. These foods of milk origin create intestinal allergy and diarrhoea.

Fat Deficiency

Fat is a high density nutrient in the feed. Fatty acids like linoleic acid and linolenic acid are prepared from fat nutrient. Linoleic is an essential fatty acid. These fatty acids cannot be synthesized by the birds. It is necessary to check exact requirement of the fatty acids in the feed. This fatty acid is basically required for cell wall structure, maintenance and for production of prostaglandins. Deficiency of fatty acids in the chicks causes poor growth and accumulation of fat in the liver. Lack of fatty acids in layer birds induces production drop and drop in hatchability.

There are two types of fatty acid's such as saturated and unsaturated. Unsaturated fatty acids during chemical changes create many damages but the process of destruction of fatty acids, inactivation of amino acids, effect on activity of vitamin A, D, E and biotin are some of the unwanted changes. Vitamin A controls regular supply of fatty acids.

Vitamin Deficiency

Vitamins comprise of many parts. Vitamins are classified as fat soluble and water soluble with the exception of vitamin C. All vitamins are required for physiological function and metabolic activity. Deficiency syndrome needs to be identified in a correct way. For example, perosis in birds is caused mainly by deficiency of choline, nicotinic acid, pyridoxine, biotin and folic acid. Poultry feed are composed by incorporating all required nutrients in right proportion. Some factors like processing of feed, transport, storage, change of feed ingredients and other environmental factors alter the nutritional balance and elicit deficiency syndrome.

Nutrient requirement of breeder ration needs to be in correct proportion. Nutritional status of the breeder ration changes the nutritional requirement of the chicks. Vitamins like riboflavin passes to the chicks through egg from parent breeder stock. Nutritional requirement of parent stock vary with age, breed and type.

Availability of the nutrients in the chicks depends on the nutrient content of the feed. Considering the solubility, the vitamins are classified into two types. They are fat soluble and water soluble. Bile salts make the availability of fat soluble vitamins to the body. Lack of availability of nutrients from the feed to the body also causes nutritional deficiency syndrome

Water soluble vitamins involves in the metabolic processes. Status of vitamins is in right progress. For example, excess choline in blood lowers the biotin content. Choline also decreases the other vitamin content in feed during storage. Higher content of protein demands more levels of pyridoxine, folic acid and biotin. Lower fat and higher carbohydrate levels in feed increases the requirements of thiamine and biotin.

Nutrients are responsible for resistance to the disease in the body. Vitamin A and E levels increases the disease resistance level. Vitamin C is able to neutralise the stress. The deficiency syndrome of each of the vitamins are discussed below.

1. Vitamin A (Anti-infective vitamin, Anti-xerophthalmic vitamin)

This vitamin is basically required for survivability and health of the body. Vitamin A can be prepared from the precursor carotene. Standardised vitamin A must be made available to the birds continuously. Synthesis of polysaccharide for cell wall protein with mannose and galactose requires vitamin A. Vitamin A should not be present in more level. Structure and maintenance of epithelial cells basically requires vitamin A. Deficiency of Vitamin A disturbs the cell structure and makes the cell easily affected by infection. Lungs, intestine and kidneys are also get affected. Affection of kidney leads to the production of folic acid which enters blood and deposited in liver and heart and also in peripheral parts like hock joints leading to paralysis.

Deficiency also accounts for drop in production, drop in hatchability, feed efficiency, lusterless feather, affections in the tissue layers and nervous system, Pathognomonic lesions of pin like pustules in the mouth and esophagus are seen early. Vitamin A accumulates in the liver. Clutch period prolongs. Nasal and eye discharges are seen.

Eyelashes are stinged and closed, purulent discharge in eyes resulting in blindness in serious affections.

Haemorrhages are quite evident in eggs. Reproduction capacity is affected. In the cocks, the sperm counts are lowered with decrease in motility rate. Morphology of sperm is also affected. Secretary parts of the cells are blocked and enlarged in size.

Affected birds can be treated easily. Inclusion of vitamin A at the rate of 5 to 10 I.U. per kg of feed lower the disease in affected birds.

2. Vitamin D (Antiracheitic Vitamin)

Vitamin D is required for assimilation of calcium in the body. This vitamin is necessary for incorporating calcium into the protein and also for phosphorus intake. Deficiency of this vitamin leads to disturbance in the deposition of calcium in the bones. This condition in chicks is known as rickets and in grown up birds as osteomalacia.

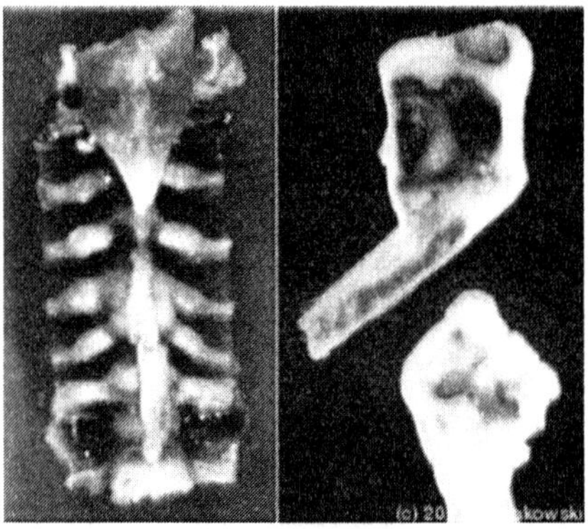

Vitamin D3 deficiency, Ca/P imbalance

Either reduced levels of calcium, phosphorus and vitamin D or imbalance ratio causes swelling of the joints and paralytic condition. Head portion of the long bones are affected. Terminal parts and joints of ribs are swollen. Soft bones parts like beak and nails are also affected.

In layers, drop in egg production, thin shelled egg shell less egg and poor hatchability are observed. It also creates a condition called osteomalacia, where in bone hardness is reduced. Bones become porous and easily breakable. Exact requirement of vitamin must be made available to the birds. Generally, vitamin A and D are two critical vitamins which are to be completely mixed in compound feed.

In the breeder flock, the requirement of vitamin D is to be met in correct proportion otherwise egg shell affection, drop in hatchability and softness of the bones are observed. Sometimes bones become structureless.

3. Vitamin E

Deficiency syndrome of vitamin E is termed as encephalomalacia. It also creates disturbances in the muscle function. Vitamin E is generally required for embryonic growth

Vitamin E primarily controls the loss of activity of fatty acids, Vitamin A, carotene and xanthophylls by oxidation process. Selenium at a level of 0.04 to 0.1 ppm compensates the vitamin E requirements. Vitamin E is important in functional activity of reproduction, hatchability and sperm production.

Encephalomalacia happens due to involvement of nervous system. Nervous weakness, torticolis, in coordination of legs with paralysis and shrinkage of muscle, leading to death of birds. Paralysis of leg and feathers are not seen completely. This deficiency occurs in 2 to 4 weeks old chicken. Sometimes attack may be seen in the age group of eight weeks. Affected birds show brain involvement in different parts.

Exudative diathesis is a condition wherein water accumulated under the skin; Blood is oozing from the blood vessels. Haemorrhages in the chest, legs and abdomen show edematous condition. Accumulation of water in pericardial sac causes edema.

During the deficiency of vitamin E and Sulphur containing amino acids, the musculature of breast region are separated. Vitamin E deficiency also affects similar changes in other muscle regions. Cereals, oil cake and green trees contain higher amount of vitamin E.

4. Vitamin K

Vitamin K is necessary for blood coagulation for the formation of prothrombin. Continuous deficiency for three weeks shows the symptoms. Sulphaquinolone inactivates vitamin K. Haemorrhages in the breast, leg and abdomen region are seen. Anaemia in chicks is common. Vitamin K is necessary to prevent haemorrhages.

Water Soluble Vitamins

Thiamine or vitamin B1

Intermediate agent is required for biochemical activity of carbohydrate metabolism. Deficiency symptoms are inappetance, nerve weakness and mortality. Symptoms occur in adult birds slowly but in chicks immediately. Initially inappetance, reduction in weight, ruffled feathers, paralysis of legs and finally birds are in incoordinated condition. Muscles are weaken and inactive resulting in lying inside with head looking upwards. This condition is termed as star gazing. Body temperature and breathing rate are reduced. Heart is also affected, birds immediately recovers by supplementation of this vitamin.

Riboflavin

This vitamin is necessary for enzymes involved in biochemical reactions of the body. It is affected by light. It is present in all the cells. Metabolic activity of carbohydrate, fat and amino acids necessarily requires this vitamin. Riboflavin is an essential vitamin which is required for growth, health and reproduction. Higher level of vitamin is seen in foods of animal origin. In adult birds, this vitamin is synthesized in large intestine but is excreted in the faeces without utilization. It cannot be stored in body. It is necessary to keep available this vitamin daily. Excess vitamin excretes through urine.

Aflatoxin prevents assimilation of this vitamin in the body. Required amounts of protein is necessary to carry this vitamin to all parts of body is another reason.

Deficiency symptoms are poor growth, inflammation of skin and nervous problem. Hatchability is affected in breeder flock. Drop in egg production, retarded embryo growth, mortality, edema and reduced weight are observed in newly hatched chicks. Lusterless feather are also observed in the deficiency condition. Curled toe paralysis is the characteristic symptoms, wherein toes in the legs are bent inwards. In addition, clubbed down condition is seen in the embryo. In this condition, front part of wings are attacking each other. Deficient chicks respond to the medicines immediately.

Pyridoxine

The source of this vitamin is meat, egg yolk and cereals. Enzyme in the ten chemical reaction of the metabolism requires this vitamin. Light inactivates this vitamin. Specific deficiency symptoms are not observed. General symptoms like drop in production, retarded growth, inappetance. Skin and feather affection are also observed.

Niacin

The vitamin is expressed as nicotinic acid and nicotinamide. It is responsible for physiological activity of enzymes required for utilisation of carbohydrate, protein and fat.

Deficiency in birds results in ruffled feathers, poor growth, changes in skin, inflammation of intestines and nervous signs. Wing bones are affected. Manganese, zinc, choline, biotin, folic acid, pyridoxine are in combination creates paralytic condition called perosis. Due to effect on growth of long bones, the growth is imbalanced. Small intestine contains inflamed sores. Reduction in lymphocyte count is also possible.

Panthothenic Acid

Small pustules are seen in the beak corners. Lacrimation and sticky eye lashes are seen. Skin in the legs get peeled off with ruptures. It is required for embryo growth and hatchability. Peak embryo mortality varies with the level of vitamin in the feed. This vitamin is essential for activation of coenzyme A. It is difficult to differentiate with biotin deficiency. Inflammation of skin, broken feathers, perosis, retarded growth and mortality are other symptoms of pantothenoacid deficiency.

Choline

Choline is essential for structural integrity. Choline is synthesized in the liver. This vitamin requires in higher level for physiological activity of the body. It is also termed as vitamin 4 or vitamin 7. It plays a vital role in metabolic process of fat. Deficiency symptoms are poor growth, accumulation of fat in liver- Fatty liver syndrome. There is a drop in hatchability and paralysis. This vitamin helps in the synthesis of amino acids like methionine and cystine.

Biotin

This vitamin is responsible for metabolic process at cellular level. This vitamin is required for growth, digestion, tissue structure and integrity, bone growth and reproduction. Vitamin is needed for enzyme involved in metabolism of carbohydrate and fat.

In the deficiency condition, there is inflammation of skin in the region of beak, eyes and legs. In growing birds, growth of long bones are affected leading to a condition called perosis. They are essential for embryonic growth. Hatchability is affected. Bone growth in legs, beak and shoulder region are affected. Fat gets accumulated in kidney and liver. Finally skin is affected. Poor feather growth and inflammatory sores in the places of joint of beak, eyes and legs are visible. This vitamin is necessary for physiological activity and for growth of feathers, nail, body growth and in egg production. Quality of chicks and good hatchability requires this vitamin for normal growth. Biotin is synthesised in the large intestines and start excreted in the faecal dropping. In order to get this synthesized vitamin, the birds should recycle the dropping for assimilation through the small intestine. Still it fulfills only 10% of total need. Even the required levels of vitamins is present in feed, it does not necessarily available as such to the birds due to various factors. Entire vitamin present in maize is utilised. But in wheat, only 5% of vitamin is made available. Due to various chemicals present in the birds, availability of this vitamin to the birds is differed. Presence of avidin in the raw eggs, mycotoxins, certain types of toxins and diseases prevent the available biotin in the feed to the birds.

In general, deficiency of this vitamin cause skin and bone affection in chicks. Fatty liver and kidney syndrome is also evident. Affection varies with the enzymes involved in the physiological activity of the body. It also differs with carbohydrate and protein levels.

Folic Acid

This vitamin is present in all the cells and is present in high levels in green leaves. It is essential for RBC synthesis. Easily destroyed by environmental temperature. Its deficiency causes poor growth, inappetance, diarrhoea, anaemic condition. Improper feather growth and paralysis are also evident. In the breeder flocks, poor hatchability and continuous embryonic mortality are seen. Leg growth in the embryos is also affected. Choline is replaced to certain extent by this vitamin.

Cyanocobalamine

This vitamin is critical for involving the carbohydrate, protein and fats in the metabolic process. In the deficiency condition, blood circulation is affected leading to production drops. Effect in embryo growth, embryo mortality and poor hatchability are common symptoms. Anaemia and nerve weakness are observed. Vitamin B_{12} content is high in animal feed sources. This vitamin is necessary for synthesis of methionine and serine from choline and homocystine and also serine from glycine.

Fish and fish products are the good source of this vitamin. It increases fat level in chicks. Feed should contain high level of vitamin B_{12} other wise, it affects chicks growth.

Calcium and Phosphorus

In poultry, 99 percent calcium and 80 percent phosphorus are present in bones. Calcium exists in the state of calcium hydro phosphate which gives hardness to the bones. In the egg shell, calcium carbonate is an important mineral. Blood coagulation activity of muscle, nerve muscle interaction, triggering of nerves, enzymes and mineral status in the body demands calcium. Nucleic acids, phospholipid, phosphoprotein, enzymes, acid base balance, chemical reaction, synthesis of carbohydrate and protein by metabolic process involves phosphorus and other minerals.

Calcium and phosphorus must be made available in feed easily. Calcium content in animal protein is higher. Phosphorus in vegetable feed source is not available easily. Vitamin D is responsible for utilisation of calcium in the feed. It is important to maintain the ratio of calcium, phosphorus and vitamin D for proper utilisation.

Decrease in levels of calcium and phosphorus create the problem depending upon the age groups. Deficiency in grown up birds causes osteomalacia and in the chicks it results in rickets.

Lower levels of calcium decreases physiological activity of the body and induce cannibalism in birds. In the layers, egg shell is thin and sometimes without shell. In the birds, the calcium requirement varies with egg production. In addition to right requirement, the calcium must be in available form. It is due that in the course of feed mixing and feed distribution, calcium may not uniformly displaced in the feed.

Higher levels of calcium affect the kidneys. There are two types of bones. First type of bones come under hard bones category and second type represent soft bones. In soft bones only, the calcium is stored.

Laying birds get the required amount of calcium from the soft bones. At the time of drop in calcium level, the daily requirement is derived from the hard bones. Due to this, the first type of bones become less hardness and are likely to be broken easily; bending of spinal cord and paralysis, due to nervous involvement are the common incidences.

Sodium Chloride

Sodium chloride as a salt of taste is responsible for maintenance of osmotic pressure and acid base balance. Sodium keeps the intracellular fluid in liquid condition and stimulates the nervous system. Chloride is present in both intra cellular and extra cellular fluids. Chloride from part of hydrochloric acid is secreted from stomach. Deficiency of this element induces many changes. Retarded growth, drop in egg production, dehydration, loss of activity of muscle and nerves are some of the common deficiency symptoms.

Common salt in poultry feed fulfils the requirement of sodium content in the feed neither go beyond 0.15% and not lower than 0.12%. Increase in salt content in feed leads to watery droppings and ascites in chicks. Secondary problems like increase in moisture content of litter and poor growth are evident due to watery droppings. Further inflammation of legs, joints and breast reduces the excess quality.

Sodium, potassium and chloride level in layer birds must be in correct proportion. Good quality and hard egg shell requires bicarbonate. Bicarbonate elements are getting available to the birds from sodium bicarbonate. Hence requirement of sodium in birds needs to be accounted based on sodium chloride and sodium bicarbonate level. Totally, salt content in the feed should not exceed 0.3%. Reduction in level of sodium chloride creates many problems in layers. Lack of sodium chloride stops the egg production completely. Lower level reduces the egg production and make the birds to move then and there. Cannibalism is in higher rate. Birds due to cannibalism cause injury in the cloacal region.

Manganese

It forms an important element in enzymes and is the basic element in gene molecules. This element is necessary for photosynthesis in plants and respiration in the birds. These elements are not readily available in birds. The mineral is present in correct quantity in livestock. But in birds this element appears in lower level. This is due to many reasons.

The requirement of this element in poultry is high. Even it is present, the element is absorbed in the intestine. Presence of calcium and phosphorus to certain extent inhibit manganese availability.

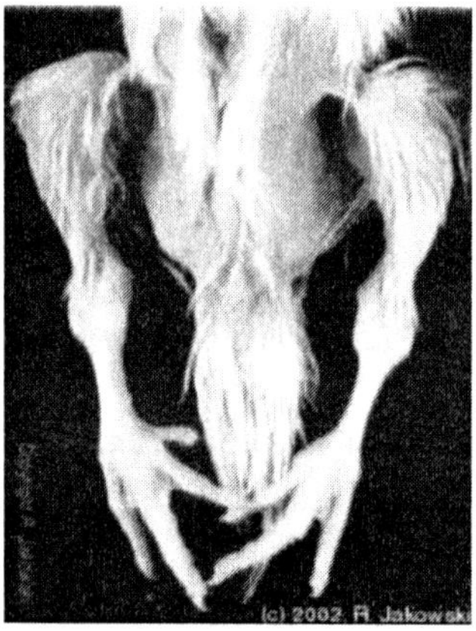

Perosis - Mn, choline, biotin, folic acid, B6

Manganese forms a basic component of the enzymes involved in the biochemical reactions of the carbohydrate. Beside, formation of egg shell and growth of bones requires this element. During the time of deficiency, structure of bones is affected and egg shell become thin. In growing birds, the mineral is required for uniform growth of bone. This element in combination with other nutrients like choline, biotin, niacin, folic acid, zinc and pyridoxine prevents the incidences of perosis in birds.

Chapter-36

Vaccines and Vaccination

Modern poultry industry deliberately depends on vaccine for control of diseases and loss of productivity. Vaccine is one of the important reasons for rapid and fast growth of the poultry industry. Even with sound biosecurity measures, it is very difficult to control many infections which are ubiquitous, worldwide and airborne. Nowadays main approach for control of diseases is by the usage of vaccines.

Primary dose of vaccination is to reduce the losses due to morbidity and mortality due to infectious agents. It is very important in layer to prevent drop in production and lowering of egg quality. Vaccination of breeder flock had a twin advantage of either by preventing the vertical transmission of diseases or by keeping maternal antibodies to prevent outbreak of diseases. Vaccination against salmonellosis guards human being from the possible zoonotic infection. Even though vaccination give 100 per cent protection against infectious diseases, vaccination is definite need to keep up health of the flock and for biosecurity measures.

Nature of Vaccine

At present, two main types of vaccines are available for poultry, which are live and killed. Single vaccines, containing multiple protections against number of bacterial and viral diseases are also available now.

Killed vaccine usually consists of inactivated antigen along with adjuvant such as oil emulsion or aluminium hydroxide. These types of vaccines usually give higher and prolonged immunity, especially after the usage of live vaccines as priming. These types of vaccines are usually multivalent and should be vaccinated individually.

On the other hand, live vaccines can be given in different routes like drinking water or aerosol or eye drops or by injection. It usually contains one type of antigen only. Live vaccines are prepared usually by making disease organisms less virulent or by using mild virulent strains. In case of live vaccines, quantity of antigen required is less, because the vaccine antigen is capable of multiplying the living bird system. The target organ for multiplication is respiratory tract for IRT and IB, or the intestine for AE and the bursa of fabricius for IBD. Live vaccines are established to the local or mucosal immunity or systemic or general immunity. Critical factor for inducing good immunity is the multiplication of the organism in the vaccinated birds and further excretion of the vaccine antigen is helpful in creating flock immunity. This type of flock immunity by cycling of antigen is advantageous for diseases like IBD, ND and IB. Lateral spread of virus is not favourable for the disease like AE or IB. Such lateral transmission of these viruses in older birds can activate production problems. Sometimes vaccine reaction may be observed like coughing or snicking after ND vaccination, which show vaccine is responded in the system. This vaccine reaction will disappear within few days, unless complicated by existent mycoplasma or bacterial infection.

When designing the vaccination programme, it is advisable to take into account various factors like,

- Immunological status
- Commercial factors
- General health of the flock
- Genetic constitution
- Cost benefit of vaccination
- Duration of immunity
- Antibody status including maternal antibody.

After formulating the vaccination schedule, method and frequency of vaccination need to be considered. Vaccination can't be uniform.

Route of Administration

Live vaccines are usually prepared in vials as freeze dried form. Storage temperature must be 4-8^{0}C. It should be protected from heat and light. Cell associated MD vaccine is preserved in the liquid nitrogen. After reconstitution of vaccine, it contains cells having MD virus, which is termed as wet vaccine.

Drinking Water

Vaccines should be reconstituted in clean cold water having skim milk powder in the ratio of 2:100. The skim milk powder have to be mixed in the water 30 minutes before mixing of water, giving a chance for neutralizing of any damaging components in the water such as chlorine or metallic ions. It is advisable not to mix the vaccine in metal storage tanks. Main caution is that drinking water to be provided should be wholesome. This water at the time of administration of vaccine should not contain any water sanitizers at any level. Plastic header tanks are recommended for administration of vaccine through drinking water. The vaccine should be used as soon as possible at the maximum time at 2 hours. Water supplying system should be thoroughly cleaned including filters or check values.

Aerosol

This type of administration enable for mass vaccination. The particle size of 5mm is recommended at bird level, so that these particles can penetrate deeply into the respiratory tract. Sometimes this particle size can induce the vaccine reaction and complication with E-coli infection. Now warn spray of 100mm diameter is recommended to avoid untoward reaction. This type of administration is mainly used for RD vaccine which can be mixed either with distilled or de-ionized water for greater efficacy.

Eye Drop

Even though this method of vaccination is easier but time consuming. However, it is an effective method. Equal care is necessary for giving correct dosage level. This type of vaccination method is most suitable for IRT and RD.

Infection

MD vaccine, chicken anaemia virus vaccine and reo vaccine and RD vaccine virus are usually administrated by injection. This is the only route of administration for killed vaccines. Automatic syringes are used for vaccination purpose. Regular checking is important to monitor correct dosage. Needle should be changed frequently for controlling contamination. Intra muscular injection are usually made in breast or leg and subcutaneous injection in the neck region.

Wing Web

This type of administration is prepared for pigeon pox vaccine. Vaccine should not come into contact with eye or mouth. After vaccination, the bird should be checked for post vaccination reaction to attain a take.

In feed

Thermo stable RD vaccine is possible to administrate through feed. This type of approach is recommended for backyard type of farming.

Through Ova

This vaccination method is tried for mareks disease, wherein the vaccine is administrated at 18th day incubated egg.

□□□

Chapter-37

Disinfectants and Disinfestants

Health of the flock is primarily depends on the adoption of effective disinfectant procedure. Selection of right disinfectant and adoption of appropriate disinfectant methods decides the microbial flora density of the farm. Sunlight and heat are commonly used as natural disinfectants. But, however due to resistance of the certain organisms and also increasing farm flock density nowadays mostly resorts to chemical disinfectant.

Disinfectant is an agent or substance that disinfect chiefly by eliminating or killing pathogenic agents or making them inactive. Disinfection is the process or mechanism of destruction of pathogenic organism. Likewise sanitizer is to reduce and control microbial organisms from active multiplication.

Properties

Ideal disinfectant should have favorable characteristics like low cost, ready solubility in hard water, relatively safety for human stability, compatibility during usage, free from obnoxious odour, less toxic, broad spectrum and residue free. Effectiveness of the disinfectant depends on the contact between surface and agent, duly free from debris and organic matter.

Phenol (Carbolic acid)

Phenol is a chemical substance derived from coal tar. It is a colourless crystal having characteristics familiar odor. It is a costlier preparation, due to which it is difficult to use as common disinfectants. This is used as base to estimate phenol co-efficient of various disinfectants. Some market preparation has residual activity even after drying, with advantage of suppression of microbial population.

Cresols

It is dark yellow or brown liquid easily miscible with water. It is extracted from coal tar and chemically related to phenol. Commercial brands are available with soap.

Bisphenols

Bisphenols are having two modified phenolic molecules. Combination of chlorine and bisphenol are much effective. Chlorophenols are having good antifungal activity. Combination of bisphenols is used as disinfectants.

Pine Oil

It is having pleasant colour with good disinfecting action. It is less injurious to skin. It is insoluble in water, hence used as emulsion form.

Hypochlorite and Chlorinated Lime

Active principle is chlorine. Hypochlorite is having 70% available chlorine. Hypochlorite contains calcium hypochlorite and sodium hypochlorite (Naocl) in combination with hydrated trisodium phosphate and as liquids having Naocl chlorinated lime (bleaching powder) prepared by saturating slaked lime with chlorine gas, which is supposed to be one of the best disinfectant. Sodium hypochlorite is available in liquid at a concentration of 1 to 15 %. Germicidal activity of the hypochlorite depends on factors like available chlorine and P^{H}, which are responsible for formation of hypochlorous acid. Germicidal activity of the disinfectant depends on the P^{H}. In lower P^{H}, better is the activity. Increasing temperature enhances the germicidal activity. When used as per direction, it is very effective disinfectant. This disinfectant is best applied in hatchery equipment and accessories.

This can be used in cement surfaces after thorough cleaning for better efficiency. Stock solution should be kept in dark, cool place and tightly sealed. It is advisable to go for fresh solution for better efficacy. These types of products should be handled cautiously to prevent damages to debris, leather and metal.

Organic Iodine

Iodine as such is one of the effective disinfecting agents. Now, iodine is used in combination with organic complexes referred as tamed iodine. Iodophore is a combination of iodine with a solubilizing agent, favoring slow release of iodine when mixed with water. This type of disinfectant refers to products available using iodine complex and surfactants. This combination increases the germicidal activity of the agent. Such combination also avoid certain disadvantagous like toxicity, irritation, foul smelling and staining. It also favours easy solubility. The presence of the detergent makes the compound stable. There is no foul smelling in this disinfectant. Wide numbers of commercial preparations are available with broad germicidal activity. Many of the disinfectants have germicidal indicator like colour variation. As the germicidal activity lowers, amber colour fades to colorless. The product can be used in cold or hard water. This can be very well used in the inanimate objects. In clean areas, the disinfectant is having very good effect.

Quicklime

Quicklime when come in contact with water liberates heat and oxygen, which is responsible for disinfection action. Chemically this disinfectant is caustic in nature and should be used cautiously.

Formaldehyde

It is gaseous in nature and available as 40% solution with base as water. It is marketed as formalin. Even in powder form it is sold as Para formaldehyde. However, very often, addition of formalin to potassium permanganate (KMNO4) liberates formaldehyde gas. It is advisable to use earthen or metal container and not glass, since in the course of reaction heat is liberated. The ratio of two volume of formalin with one dry weight of $Kmno_4$ (1g $Kmno_4$:2 ml formalin) is recommended. If high amount of formalin is used, it is wasted unnecessarily. On the other hand, too high amount of $Kmno_4$ is used;

it is unchanged and is wasted. Potassium permanganate is toxic in nature. Both these chemicals have to be stored in safe containers.

Formaldehyde gas has certain disadvantages like volatile, pungent odour, caustic action and tendency to harden the skin. It is extremely irritant to conjunctiva and mucous membrane. It is mostly used for fumigation of hatching eggs. Even in the presence of organic matter, it acts well. 30% ammonium hydroxide can be used as neutralizing agent for formaldehyde, but half the quantity of formalin can only be used. General recommendation of hatching eggs are 60 g $Kmno_4$:120 ml formalin/100 ft^3 (2.8 m^3) cabinet space at 21.1°C and 70 % humidity with 20 minutes fumigation time. Higher the humidity and fumigation, more effective the fumigation. As to have better penetration of formaldehyde gas, it is recommended to increase quantity of chemicals (75g $Kmno_4$):150 ml formalin/100 ft^3) with higher temperature of 32.2°C, relative humidity of 90% and contact time of 30 minutes. Hatching machine can be fumigated inside and within18 days of incubation.

Antifungal Imidazole

Imidazole or enilconazole is the effective antifungal agent. It acts by interfering with ergosterol synthesis at low concentration as fungi static and by causing membrane damage at higher concentration as fungicidal. These antifungal agents should be combined with antibacterial disinfectants for complete broad spectrum hatchery sanitation.

Copper Sulphate

Copper sulphate is having marked toxic effects on lower form of life. However, it is not widely used as disinfectant. It is also toxic to algae and fungi. In chicken, it is used at a concentration of 0.5 to 1% in feed for shorter periods without recorded toxicity. Water containing 1:2000 concentrations can be used in chicken. Turkey will refuse water containing copper sulphate. 0.5% concentration of CuSo4 is recommended as disinfectant.

Quarterenary Ammonium

This type of disinfectant are non corrosive, clear solution, odourless, non irritatant, deodorant and cationic. It is highly stable and non toxic. Before usage, thorough cleaning of the surface and removal of anionic detergent and soap is essential to have good sterilizing purposes.

Sunlight

Sunrays are having good disinfecting property. But to have better germicidal effect, the surface should be exposed thin layer to dirty sunlight. As such it is not practicable method. Surface cleaning is necessary to have better effect by itself.

Hot Water

Boiling water or live stream is having good disinfecting action. Addition of disinfectant increase the cleaning .Live stream needs to be applied directly and closely for disinfection action.

Dry Heat

Dry heat is very effective killer of pathogenic organism. Caution should be exercised in usage of direct flame to avoid fire hazards. It is recommended on cement surfaces. Direct flames can be used to eliminate feathers and fluff accumulation.

Disinfestants

These are used to destroy ectoparasites like lice, mite, ticks and fleas. Most of the pesticides are highly toxic to humans and livestock. Many disinfestants are destructive to ectoparasites when they come in contact with directly whereas most pesticides are not having disinfectant properties.

Disinfestants are available in the form of crude oil, 40% solution of nicotine sulphate, fog or mist insecticides, and drugs of systemic inhibition and as dusts and sprays.

❑❑❑

Chapter-38

Hatchery Sanitation

Hatchability is important in hatchery operation. Difference of 5% hatchability drastically affects the profit ratio in hatchery. Many factors govern the hatchability performance of chicks. Some of the major factors which decide the hatching performances are:

1. Fertility of breeding female flock.
2. Nutrient status of the breeding flock.
3. Genetic status of embryo.
4. Selection of hatching eggs.
5. Hatchery management.
6. Infectious organisms in egg.
7. Vertically transmissible diseases in breeding flocks.

Ways and means to control entry of infectious organisms in egg reflects directly on hatching performance. Infection at farm and egg storage place should be avoided without concentrating on these efforts, the use of disinfectant and disease control measures cannot be useful. It is necessary to know the entry of infectious organisms in egg.

1. Infectious organisms in the hatching egg are possible due to unhygienic conditions. Bacterial organisms are not present in egg white and yolk immediately after laying. But afterwards by many ways, infectious organisms enter the egg. Ovarian infections in the parent stock deliberately contaminate the egg.
2. From the release of egg yolk to egg laying, infectious organisms if present enters the egg by many ways. It actually happens due to infection of ovary and reproductive tract.
3. After laying due to unhygienic conditions, the infectious organism present in litter and environment enters the egg.
4. Infection at the place of egg storage and from the egg transporting vehicle is feasible. Egg transporting carts and boxes also transmit the disease. Psycrophillic organisms grow in the egg in egg storage place.
5. Diseases are also transmitted from the hatchery. Hatchery sanitation is most important. As other wise it is conducive for the production and multiplication of all types of organisms. Such organism also enters the egg easily. Exploding eggs also affects the other good eggs.

Temperature and relative humidity maintained in the egg incubators and hatchers are so conducive for growing of pathogenic microbial organisms. In these ways, entry of microbes is possible in the egg.

Embryonic Mortality

Generally embryonic mortality happens at two places. First it occurs during the period between 2 and 4 days and second during the period between 18 and 21 days of incubation. Early embryonic mortality is mostly due to parent stock. Improper hatchery sanitation accounts for late embryonic mortality. Microbial organism enters through small pores in the egg. Cold storage condition shrinks the membranes surrounding the egg, leads to increase in size of egg pores. Inflammation of ovary makes the egg easily prone for infection.

Egg yolk is a place for harboring the infection. Presence of lysozyme destroys the entry of organisms. Some of the reasons for embryonic mortality are:

1. E-coli type of organisms are mostly responsible for mortality.
2. Transovarian infections like Salmonella and Arizona causes mortality.
3. Exploding eggs are seen in hatchery. Such egg exploders burst huge quantity of microbes which infect the other eggs.
4. Avian encephalomyelitis induces mortality at late embryonic stage.
5. Presence of aflatoxin is a factor for mortality during first extent.
6. Infections like *Mycoplasma* and *Staphylococcus* also causes embryonic death.

Even then hygienic surrounding and proper hatchery sanitation entirely controls the embryonic mortality.

Disease Transmitted from Hatchery

1. Unabsorbed yolk: Depending upon the egg hatching and brooding place, this type of diseases occur. Unhygienic surroundings and failure to clean the hatchers using formaldehydes are responsible for such type of diseases. Infectious organisms enter the body through the naval cord. Improper healing of naval cord causes enlargement of naval. Swelling of stomach and chest are also seen. 70% of this type of disease is due to E-coli. Presence of infectious organisms in the droppings also cause infections.

   During the course of infection, mortality occurs during the first three weeks. Some strains of E-coli causes 100% mortality. More than four days of infection induces inflammation of heart and foul smelling in the naval cord. Sometimes, there may not be mortality but severe losses in weight gain could be seen.

2. Pullorum disease: It is another important disease. Infected flock encounters much mortality. Drop in hatchability also disturb the economic status of the hatchery. It is one of the important diseases transmitted by trans-ovarian transmission. These organisms are responsible for inflammation of the ovary and reproductive tract where these organisms are physically present.

   As it is transmitted vertically, such type of disease should not be present in breeding flock on any account. Screening of the breeding flock for this type of infections is quite appropriate. In addition, the infection is spread from infected flock to healthy flock. Feathers and dusts present in the hatchery also transmit the infection.

Isolation of organisms is possible from the infected birds. Blood examination is necessary for confirming the disease. Systematically the breeding flocks should be subjected to blood examination at 10th and 20th week before going into egg production.

3. Fowl typhoid: It is caused by *Salmonella pullorum*. This disease is easily transmissible from hatchery. Affected birds' shows higher incidence of mortality. In grown up layer birds, the infection reduces the egg production. Decomposed fish meal, contains this type of organisms. It is easily identifiable by laboratory blood examination.
4. Avian mycoplasmosis: *Mycoplasma gallisepticum* is causative organism for avian mycoplasmosis. In the turkey, it causes sinusitis. Some times combination of E-coli, viral disease and avian mycoplasmosis causes heavy mortality. It is transmissible from parent stock to chicks through egg. As to avoid trans-ovarian infection, the breeding flock should be screened for this type of infection before egg production. Healthy and disease free birds should be used for the hatching egg production.
5. Ranikhet Disease: Infection in the breeding layers disturbs the egg number, egg shell thickness and pore size. Owing to this, higher water content is released leading to a drop in hatchability. These organisms are isolated from egg yolk, infertile egg, dead embryo, and four day old embryo.
6. E.D.S.76: It is caused by a virus. Drop in production, uneven shape of egg, change of colour and thin shell egg are important signs of this disease. Lower hatching performances in such infected eggs are usually seen. Drop in livability rate in chicks are also another important factor commonly observed in this infections.
7. Avian Encephalomyelitis: Eggs collected from infected flock result in mortality during the last three days of hatching. Drop in production corresponds with drop in hatchability. Chicks hatched from such eggs are debilitated.

Diseases of respiratory tract, infectious coryza and infectious bronchitis also affect the egg production and hatchability. Drop in egg quality ends with drop in chick's quality. As such, hatchery sanitation is a critical factor. If the parent flock is healthy, the 100 chicks received from this parent birds are also healthy.

□□□

Chapter-39

Quality of Chicks

Quality of the chicks is the critical factor for efficient production and profit. Consequent to increase in demand of chicks, there is corresponding increase in hatcheries. Increase in demand of chicks result in decrease in quality. Quality of chicken lowers due to many reasons. Poor quality predisposes many problems in the farm.

Primarily good hatchery and breeding flock are important for quality of chicks. Much attention in the management of breeding flock, setter and hatcher sanitation are necessary for production of good quality chicks. Parent stock should be free from contagious disease problems. Various factors responsible for production of quality chicks are.

1. Genetic potential of breeding flock

Inherited genetic character of breeding flock is important. Some breeds are capable of producing higher egg weight. Good egg weight inturn produces reasonable weight of chicks. Chicks of recommended weight are thriving very well in the environment. Quality of the chicks coincides with weight of the chicks. Contrarily, egg number and egg weight are acting indirectly. Increase in number of eggs produced result in drop in egg weight. Lesser egg weight hatches out lesser weight chicks, resulting in drop in quality of chicks. These lower weight range chicks are easily susceptible to the diseases having less production performance. It is not advisable, just for the cause of increase in production of chicks; the egg number should not be increased. Breeding flock should possess very good disease resistance and much better production potential. Genetic character, environment and feed are also the governing factors for increasing the egg weight.

2. Nutrients

Production performance of any birds directly related to quality of feed. Quality of the chicks depends on the nutrient status of the birds. Recommended levels of protein and amino acids for breeding flocks are necessary. Calcium and phosphorus levels are also at right level. Mineral and vitamin content are critical factors determining the quality of the chicks. Quantity of feed and fatty acid compositions also reflect on the quality of the chicks.

3. Environment

Environmental surroundings for the breeding flock should be conducive and comfortable to the birds. Healthy environment and needed comforts favours normal production. Farm designs, farm microenvironment, ventilation, moisture content are to be taken into account for getting good expected production. Alteration of any of these conditions lowers the production. Litter management needs to be in proper level. Ammonia concentration should not exceed the permissible level. Intensity of light and duration of light are also accounts for egg production. Different light schedule alters egg weight and consequently chick weight.

4. Management

Farm management shall be as per scientific line. Drop in management threshold levels induces stress in birds. Breeders, feeder space, waterer number, waterer space, instrument design and litter density are interacting in better breeding flock management. There should not be any drop in management level.

5. Disease

Some of the major diseases are transmitted from parent flock to chicks through eggs. Ranikhet disease, infectious coryza, Egg drop syndrome, Mareks disease, Lymphoid leucosis, Infectious bursal disease, Salmonellosis and E-coli are the diseases affecting the reproductive tract, thereby affecting the egg production. Breeding flock should maintain high antibody level and good disease resistance.

Hatching eggs are to be stored in proper recommended environment. Temperature in egg storage room needs to be below the physiological zero of the embryo. Time log between egg storage room and incubator setting and sweating time are to be appropriate. Incubator and hatcher are to be maintained with proper hatchery sanitation measures.

Hatchery environment is another important critical factor. Hatchery temperature, relative humidity, oxygen level, carbon dioxide level are to be maintained at recommended level. After hatching, chicks should not be continued to dry in the hatcher for long tome. Continuous retention of the chicks in the hatchery accelerates evaporative loss of moisture from chicks leading to dehydration. Increased microbial load is culminating factor for unabsorbed yolk in the chicks.

Chicks after hatching should reach the brooder with 24 hours. Higher time interval than this, succumbs the chicks to dehydration due to water loss from the body. Chicks should not be transported in closed boxes and in higher density. Protected chick boxes are necessary for transport of chicks. During transport, chicks should not be subjected to hot or cold climate.

Importance of Quality Chick

Quality of chicks reflects on the production. Less weight chicks are having lower weight gaining efficiency. Poor quality chicks encounter with higher mortality rate. These poor quality chicks are easily susceptible to disease and stress. Minor changes in environment also affects the chicks, poor feed efficiency and poor carcass quality are frequently seen in poor quality chicks.

Key Points to Identity Good Quality Chicks

1. Broiler chick shall be weighed at 40 gm and layer chicks shall weight at 35 gm at day old.
2. Day old chicks should move freely and should be alert.
3. Bright eyes is another important mark in good quality chicks.
4. Feather should be well grown and developed.

5. Chicks should not have any defects.
6. Chicks frequently should consume water and feed.
7. Dropping should not adhere to the tail part of chicks.
8. Chicks belly should be normal.
9. Mortality rate in good quality chicks should not exceed 1 to 2% during first week of age.
10. Parent stock used for producing chicks must be free from all contagious diseases which are transmitted vertically.

□□□

Chapter-40

Farm Hygiene

Poultry farmers should be careful in so many things in the farm. Good quality chicks and feed are the critical factors to avoid any disease outbreak in the farm. Farm consultants usually give importance for the flock health. Altogether, good quality chicks, optimum management, good quality feed, comfortable environment and appropriate disease control measures are taken for efficient production in the farm. It is advisable to have good care before attending any work in the farm.

Farm Management

Poultry farm can be run with different systems of management. Still farm hygiene is critical in all management practices. Careless action can lead to major disease out break. For example, the dust and cob webs accumulated in the farm can act as a place for microbial multiplication and intermediate hosts. It also leads to reduction in wind flow. Further, this facilitates accumulation of the ammonia toxin. Ammonia is an irritating gas. Presence of ammonia gas irritates the bronchi and larynx of the respiratory tract. As to have good ventilation, the dust and cobwebs should be removed then and there.

Presence of moisture even causes problems. Excess moisture in any place is not advisable. Increased moisture in litter beyond 20% leads to outbreak of coccidiosis. Higher levels of moisture only make the oocyst gets sporulated and infective.

Waterers and feeders should be thoroughly cleaned and exposed to the sunlight before usage. Sunlight could be effectively used for eradication of disease causing organisms. Litter management needs critical approach. Powdery litter or caked litter is not good for the birds. Foul smelling litter should be avoided. Foul smelling litter indicates microbial multiplication, which further leads to production of ammonia. Bad quality litter needs to be replaced with fresh one. Good litter makes the birds to have a good external appearance.

Farm surrounding should have good atmosphere free from unwanted activities like urination, defaecation and waste disposal. Water stagnation should be avoided atleast 100 feet surrounding the farm. Even during rainy season it is essential to have lime powder at the entrance of the farm as disinfectant.

Drinking Water

Quality of the water is rather important, as to that of feed. Intake of water is twice that of the feed. In general, quality of the water is not getting due care in most of the farms. Hygienic surroundings from water yield to water collection point are quite essential. There should not be any contaminants in the water. Water should be tested frequently for the presence of disease causing organisms. Bore well water can be used for cleaning poultry sheds, water tanks, feeder drum and waterers. Disinfected and tested water should be made available to the birds. Water storage places should be properly covered. In uncovered area, flying birds like crow, parrot and eagle may likely to sit and drink water. During the course of taking water, these birds can contaminate the drinking water with faecal droppings, which often contains infectious organisms particularly ranikhet disease virus. In that way, it leads to the possibility for outbreak of ranikhet disease in the farm. Flying birds are resistant to ranikhet disease but act as a carrier for the disease.

Waterer should be cleaned properly. Sliminess inside waterers needs to be avoided on any account. Water should not be splashed inside the farm. Water should not be mixed with feed in feeders. Such possible mixing of water may lead to the growth of fungus and subsequent production of metabolites like mycotoxin. Even accidental mixing of water in feeder should be avoided to prevent mycotoxin formation in the feed. Even quality feed may also become mycotoxin contaminated feed due to increase in moisture content in feed which favours fungal growth and subsequent mycotoxin formation. Waterer should be double in numbers, to have possibility of sun drying before usage. Sun dried waterer is better to uphold the farm hygiene. Water in the waterer should be clean in the morning. But due to accidental addition of faecal material and litter, water gets easily contaminated. In course of further addition of water, at any account should not be done with contaminated water like above. Such careless approaches give a chance for microbial multiplication and further lead to disease outbreak. It is essential to remove and discard such contaminated water and provide fresh clean water. In most of the farm, neither care is taken to remove contaminated water nor to provide fresh clean water. Such action definitely results in drop in production or disease outbreak. The main point of clean water availability all times should be the prime attempt. Even cage waterers, should be cleaned properly. Automatic waterers should be checked inside properly for cleanliness once in fifteen day's atleast.

1. Farm Labourers Hygiene

Farm hygiene depends on the capability and involvement of the farm labourers. Farm labourers, should be healthy and active. Labourers should not indulge in unwanted activities like spitting and nose cleaning. Labourers should wear clean clothes. It is advisable to have labourer to get bath and wear fresh clothes before taking up farm activities. Entry of outsider and other farm labourers should be checked completely. Organised way of farm routine should be encouraged. Stress inducing actions like frightening the birds, feed wastage and water splashing must be avoided. Feed storage place should be moisture free and rat proof.

2. Preventive Measures

Even though it is possible to attempt all farm hygienic activities, still it is necessary to take up all disease preventive measures. Disease causing organisms can enter in so many ways. Farm should be duly protected from rat nuisance. Dead birds should be subjected to autopsy examination to find out the cause of mortality. Dead birds should be properly buried or burned. Dead birds should not be thrown away. By this way, disease transmission from one place to another can be controlled. Medication can be started unless warranted.

Farm hygiene is of primary importance. Poultry farm activities are concerned with live birds. As such every care and action is critically advisable. All action and mistake can lead to major disease outbreaks. All action should be taken to keep the health of the birds with efficient production.

❑❑❑

Chapter-41

Water Hygiene

Water is a critical nutrient for all living organisms. In the poultry farm, water plays a vital role. Water is essential for health of the birds and hygienic surroundings of the farm. As such, it is essential to keep the water clean. However, the poultry farmers are least bothered about the water.

Source of water may be different like river, ponds, well and bore well. Water yield point may be subsoil or deep soil. Quality of water depends on the water source. Quantity of minerals present in the water differentiates as soft water or hard water. Well water is usually hard water in nature. This type of water contains all kinds of contaminants. It is difficult to provide protected water to the poultry farm. Poultry farmers should have basic knowledge of water quality and water sources. It is necessary to know the water sanitation procedures.

New poultry farms must get the analysed water samples before starting the farm. Analytical report should contain mineral status and microbial load of the water samples. Interpret the water sample analysis report with veterinary surgeon.

Water hygiene is critically important. Cleanliness of the water should be maintained from water yielding point to water usage point. Water collection should be hygienic. Five feet surrounding the well should be cement plastered to avoid possible subsoil water contamination. At any account, litter disposal area should be a minimum of 100 feet away. Unused water from waterers should be a discharged properly. Waterer cleaning place should be an elevated place with proper drainage to avoid any stagnation. There should not be any leakage in water lines. Water storage places should be properly covered. If it is open, exposed water may get contaminated with flying birds faecal matter.

Water tanks are need to be cleaned once in fifteen days. Inside of the water tank needs to be scrabed well and cleaned properly to eliminate algae growth. Bleaching powder can be used to remove the sliminess. After cleaning, the water tanks should be exposed to six hours. Then, inside walls should be given lime coating. It is necessary to check the water for the presence of dead lizard and rats. Water collection must be at higher level without any leakage.

In the healthy farm, water plays an important role. Water is important for efficient production. Water stagnation with decayed materials is a suitable place for multiplication of disease causing organisms. Many diseases are spread by water. Microbial load in the farm varies with microbial load in the drinking water. Higher microbial load in the water not only cause drop in production but also gets excreted through droppings. Higher microbial load can also cause higher moisture content in the droppings. Such conditions can also cause outbreak of coccidiosis due to sporulation of oocysts, which is in infective stage. Further, litter fermentation increases the environmental temperature along with higher relative humidity in the environment. Increase salt content in water also increases the water content in the droppings. It is also stated that higher mineral content in water induces disturbances in metabolism and functional activities of kidney and liver.

Water can't be considered as a small thing. Careless approach leads to major problems. It is necessary to get protected water to all birds. Water sanitation is better done with 4 to 7 gm / 1000 litres of drinking water. Residual chlorine level needs to be maintained at 0.1 to 0.2 ppm. Residual chlorine level can be measured by chloroscope. Bleaching powder should have available chlorine content of 40%. Liquid chlorine is also available. 5% sodium hypochlorite can be treated with 25 litres of water. Following are some of the ways to protect the water hygiene.

1. Contamination of water needs to be avoided.
2. Water stagnation should be avoided.
3. Inner side of water containers should be free from sliminess.
4. Water storage drum should be properly covered.
5. Water should not be contaminated with flying bird's droppings.
6. It is advisable to use plastic or mud waterers since they are rust proof.
7. Clean water has to be provided in waterer.
8. Water needs to be provided twice daily.
9. Water splashing in the feed need to be avoided.
10. Water should not be contaminated with rat droppings.
11. Water should be chlorinated continuously.

❑❑❑

Chapter-42

Role of Water in Poultry Production

Water is one of the important nutrients in the poultry farms. Water is required for farm hygiene and birds health. Water is the cheapest but most indispensable inorganic nutrient for all living things. Water is an universal solvent, basic substance of all living cells and body fluids. It forms the largest single components of body. Water is present both intra-cellularly and extra-cellularly. About 65% of an avian egg and chicken body are made up of water, which indicates the importance of water in poultry nutrition.

Functions of Water

- *Thermo regulation*: Water helps to maintain constant body temperature. Birds must consume ample quantities of drinking water to keep body's moisture percentage constant. But more quantity of water is required to maintain a constant deep body temperature during hot weather.
- As water absorbs heat rapidly, its movement through the body is essential to remove excess body heat. Considerable amount of water leaves the body by vaporization from the lungs and airsacs;

since fowls have no sweat glands. Vaporization of water removes excess heat from the body and variations in the quantity depends on external temerature.

- Water is an universal solvent and media for various bio-chemical reactions in the body. Water forms the base for secretion of hormones, enzymes and other biochemical substances, which induce various biochemical reactions in the body.
- Water is necessary for the digestive processes. Some digestive enzymes are more effective when diluted in water and the OH- and H+ (which comes from water) ions are involved in the process of hydrolysis and digestion of protein, fat and carbohydrate. It aids in soaking the feed in the crop and act as a carrier during its passage through the alimentary tract.
- Water also helps in the absorption and assimilation of nutrients from the digestive tract.
- As a major constituent of blood, water acts as a medium for transportation of various nutrients, enzymes, hormones, antibodies etc. from one part to other parts of the body.
- Water as a medium, helps in the excretion of various waste materials from the body.

Sources of Water

Birds will get the water through three sources. viz.

- Drinking water,
- Feed and
- Metabolic water.

An adult White Leghorn hen consumes about 200-250 ml of water daily. In addition, it obtains about 10-15 ml of water present in the feed as moisture. Moreover, during the oxidation of carbohydrates, lipids and proteins in the body, the bird obtains respectively 0.6, 1.1 and 0.4 ml of metabolic water per gram. Totally an adult hen obtains and utilizes 260-320 ml of water daily.

Water Excretion

Water is eliminated mainly through the faecal matter of which about one fourth comes from the urine and three fourth from the intestinal tract. Nearly 40% of the water loss from the chicken's body is through respiration (Patrick and Schaible, 1981). The remaining water loss from the body is through eggs and semen.

On unit weight basis, the total amount of water eliminated through respiration and faecal discharges decreases as the size of the bird increases. Younger birds have less moisture in their droppings than older birds. A 1.8 kg bird at a temperature of 21°C losses about 43% of the moisture through the faecal makker, which accounts for 25% loss in 0.45 kg bird . Freshly voided poultry faeces contain high percentage of water ranging from 70% to 80%. Water loss by any route except the lungs include loss of electrolytes. Complex hormonal mechanism mediated by the hypothalamus and the adrenals control the water loss through kidney.

Water Consumption

Water is required in larger amount of all the nutrients. Chicken will consume about 2 to 2.5 grams of water for every gram of feed consumed under moderate temperature. As the temperature increases, the water intake also increases. With rise in temperature from 21°C to 35°C, the water consumption almost doubles. Laying pullets invariably consume more water than non layers. On equal weight basis, younger birds consume more water than older birds. High salt diet increases the water intake. Less palatable and brackish water reduces water intake. Birds fed with pellets and birds in cages drink more water. Birds generally reject warm and very cold water. Approximate water consumption at 25°C-30°C by chicken and turkey of different age groups are presented in table 1. Laying hen consume from 200–400 ml per day, depending upon the body weight, rate of egg production and ambient temperature.

Factors Influencing the Water Requirement

The water requirement of birds depend on the following factors:

(1) *Environmental Temperature*: As the environmental temperature increases, there is corresponding increase in the water consumption. Water consumption increases by 100% for a rise in the environmental temperature from 21°C-32°C.

(2) *Genetic* : It is observed that variations in water intake of birds ranging from 7.2% to 25.3% of the body weight per day in birds of different genetic make up.

Table 1. Water consumption by Chicken and Turkeys of different ages and rate of lay (L/1000 birds/day at 25⁰-30⁰⁾

Age (wks)/rate of lay (%)	Meat – type chicken	Egg – type chicken	Turkey
1	23	20	37
2	42	40	75
3	67	45	113
4	120	64	151
5	140	83	189
6	170	94	227
7	207	105	283
8	285	118	359
10	250	140	434
12	270	153	473
16	285	161	567
20	300	175	605
25 Male	-	-	700
Female	-	-	500
30 Male	-	-	850
Female	-	-	600
% of lay			
10	330	200	700
25	355	220	800
50	375	230	900
70	400	250	-
90	-	270	-

(3) *Kind of Birds*: Generally, laying hens need more water than chicks or growing birds, since increase in egg production also increases the water consumption.

(4) *Management*: Cage birds require more water than the birds reared on deep litter.

(5) *Feed*: Diets rich in protein of meat scrap, soyabean meal and fish meal origin increases the water requirement where as milk protein decreases the water requirement. Feeding high energy feed requires less water than feeding low energy rations. Feeds which are high in fibre increase the water requirement. Many minerals influence the water requirements of poultry of which common salt has the greatest role. Birds fed with high sodium chloride diets consume more water and also excretes more water leading to wet litter problem.

(6) *Diseases*: Many diseases of the intestinal tract increase the water intake of the birds and results in higher moisture excretion in the faeces.

Water Quality and Analysis

Water quality is used to define those chemical, physical, micro-biological or radiological characters, by which the overall acceptability of the water is evaluated. Good water supply to poultry is important, to maintain their health and high rate of production. Water contains many substances in solution or suspension, many of which affect its palatability and quality. Good water is colourless, odourless and tasteless. Before being used in the farm, a sample of water should be submitted to a laboratory for analysis of chemicals and microbial purity. Threshold levels of various chemicals in water for poultry are presented in Table 2. Quality of water should be judged from the following analytic points:

Table 2. Threshold levels of microelements in water for poultry

Microelements	Threshold concentration mg/L (ppm)
Total dissolved solids	1000 - 2500
Total alkalinity	500
Calcium	250
Sodium chloride	500 - 1000
Bicarbonate	500
Chloride	1500
Fluoride	1
Nitrate	40 - 50
Nitrite	0
Sulphate	250 - 500
Copper	1
Cadmium	5
Iron	0
Arsenic	0.5
PH	7.6 to 8
Zinc	5
Manganese	0.05
Boron	1
Uranyl	5
Pesticide (aldrin, Toxophene, DDT etc)	0.001

Colour: Any colour is due to the presence of substance like tannins, iron salts etc. Tannic acid reduces the digestibility of proteins.

Turbidity: Mostly particles in the solution cause the water to be turbid.

P^H: A p^H of 7.2-8.0 is recommended for use in poultry farms.

Hardness: Dissolved salts of calcium carbonate and magnesium form scale and sludge and cause the water to be hard. Hardness affects the taste of water. Soft water of 15 to 50 ppm is preferred for poultry.

Total solids: Total solids represent the total amount of solid material present in the form of suspension or solution.

Iron: Presence of Iron causes staining on all things when it comes in contact. It rapidly lowers egg quality.

Sulphur: The presence of sulphur can be detected by its odour (rotten eggs). Water which is high in sulphur is not palatable and tends to causes hemorrhages under the skin.

Nitrogen: Presence of nitrogen is an indicative of decaying organic material and is a measure of contamination.

Chloride: Should not be present in excess of 1000 ppm, since when excess presence affects palatability and causes diarrhoea, ascites, dehydration and mortality.

Trace Minerals: Fluorine, Iron, Selenium and Molybdenum are often present in toxic amounts. Excess fluorine affects carbohydrate metabolism. However, birds can tolerate fluorine toxicity better than cattle. Iron causes staining and deterioration of egg quality, while selenium is extremely toxic.

Bacteria: Type of bacteria present in water is more important than their number. Many infectious diseases can be transmitted through drinking water. Among these E.coli, Enterobacter and Salmonella are most important pathogenic organisms. Ducks perform well with fairly high level of E.coli contamination through water than chickens. Turkeys have a higher tolerance for nitrates than cattle.

Parasites: Coccidiosis is one of the important diseases transmitted by drinking water due to contamination with droppings. Ova and eggs of various cestode and nematode parasites are transmitted through drinking water.

Water disinfection: The water disinfection processes aims at removing or reducing the chemical and biological impurities, at an economical rate by the process of sedimentation, coagulation by boiling, filtration, sanitizing with chemicals like chlorine, iodophores, ozone etc. Water disinfection processes of current significance include one or combination of the following methods:

- Physical treatment such as storage, sedimentation, filtration and straining
- Irradiation such as ultra-violet light treatment
- Deionization
- Alum and activated carbon treatment
- Treating with alkalis and acids
- Surface active chemicals such as quaterenary ammonium compounds
- Oxidation with ozone and H_2O_2
- Treatment with halogens like bleaching powder, liquid chlorine, iodophores etc., which will destroy the microflora in the water. Recently water is treated with food grade polyphosphates; which act as sequestering agent by reacting with many inorganic chemical contaminants.

Water Management

Wholesome drinking water in sufficient quantity should be provided to all the birds. Optimum waterer space should be provided based on the age, type of birds, type and form of feed. Birds should never be deprived of water. Waterers should be cleaned periodically. Water spillage should be avoided. For this water level should be atleast 1 cm below the brim or edge of the waterers. Waterers should be distributed uniformly. In deep litter house, birds should not walk more than ten feet to reach the nearest waterer. Waterer should be keep at sufficient height so that the tip of the waterers should be on par with the back of the birds.

Water proportions should be maintained and operated properly to have effective medication and vaccination through drinking water. When medicated water is given it should be consumed preferably within two hours, without any wastage, followed by providing ordinary water. When vaccines are given in drinking water, such water should be neutralized for the possible presence of any disinfectant with skim milk in the ration of one part to 400 parts of water.

Birds in normal health consume right quantity of water. In stress and disease condition water consumption may go up or reduce, as such knowing the actual water intake of the birds may indicate the health status of the birds. So, it is advisable to install water meters in each house to know the water consumption.

While cleaning the waters it must be seen that slimeness should be removed and to be cleaned using sanitizers.

□□□

Chapter-43

Adulteration in Poultry Feeds

The quality of the feed and feed ingredients are need to be given much importance. Any commodity, as the demand increases, there is a tendency to drop in quality and rise in price. Quality of the feed determines the performance of the birds and hence whole profit margin of the industry. Owing to involvement of the mischievous traders, the possibility of adulteration inherently present or added wantedly is increasing. Adulteration in the poultry feed can be the presence of substandard feed ingredients or by inclusion of substance which affects the normal physiology of the birds.

1. Moisture content

Normally the recommended level of the moisture content in different feed ingredients should not go beyond 10%. During rainy season and by wanted addition of water in the feed with an intention to boost up weight, moisture content in the feed will increase. Since the moisture content in the feed ingredients favours the growth of fungus, there is every possibility for growth of fungus and consequent production of mycotoxins. Mycotoxins are dangerous for the health of birds and lead to drop in production and mortality. Hence, while screening the feed ingredients during procurement, greater care should be taken to see that the moisture content is within the permissible level.

2. Mycotoxins

During initial stages the higher moisture content can reflect on the weight of the feed. If this moisture containing feed is neglected to get it dry, there is every possibility for growth of fungus. Fungus like *Aspergillus flavus, Aspergillus paraciticum, Aspergillus fumigatus and Aspergillus ochraceus* can grow in the feed leads to the production of mycotoxins. It is a problem in cereals, vegetable oil cakes, fish meal and meat meal. Mycotoxin is a dangerous potential poison causing hepatitis, nephritis and generalized toxemia. Among the consignment, even if one bag is having fungal growth, it will spoil the entire mixing lot. Incorporation of one or two bags in between the stock may likely to happen. Samples should be seen in maximum possible bags. If the consignment is found to contain contaminated samples, it must be withdrawn.

3. Paddy husk

It is a common problem in the rice bran, wheat bran and rice polish. Since, paddy husk happens to be a cheaper feed ingredient, it is grinded nicely and is adulterated in bran. Digestibility value of the paddy husk is almost nil and its addition in the bran unnecessarily increases the crude fibre content of the bran. Its incorporation in the feed dilutes the nutritional density and lowers the quality of the feed. Presence of paddy husk in the bran can be easily detected by feeling the small part of the sample between fingers, in which case there is rough texture.

4. Rancidity

It is a common problem in the vegetable oil cakes, rice polish and in the feed ingredients rich in oil. If the oil rich feed ingredients are stored for long time, the oil gets rancid and this leads to production of other fatty acids. These fatty acids cause toxic hepatitis and mortality in birds. It is more common in gingelly oil cakes and the chances are low in the solvent extracted oil cakes, since the oil content is less in solvent extracted oil cakes. Its presence is easily assessed by smelling the feed samples in which spoiled oil flavour can be detected.

5. Urea

It is the most common problem in the fish meal and meat meal. Price and scarcity of the fish meal paves the way for urea adulteration. Quality of the fish meal is assessed by crude protein content. Protein present in the fish meal is of good quality and having better biological value. Its protein is of nitrogenous origin. Urea is having non protein nitrogenous substances and is not showing any biological value. Adulteration of urea in the fish meal boost up the protein content but without any biological value. Moreover, urea is toxic to the birds and causes the gout condition which produces drop in production and mortality. Hence, altogether presence of urea in the fish meal should be assessed.

6. Salt

Fish meal prepared from the sea origin contains sodium chloride salt as a normal constituent. Presence of salt in the fish meal upto the maximum level of 3% is permissible. Sometimes common salt is added to increase the weight of the fish meal consignment. When the sodium chloride is present beyond 3% in the fish meal, it naturally increases the salt level in the compound feed. In the compound feed, the chicks can tolerate sodium chloride level around 0.4%, beyond this limit it produces salt toxicity in the chicks. Salt toxicity leads to enlargement of abdomen, ascites and mortality. Hence, while procuring the fishmeal attempts should be made to screen the salt content of fish meal.

7. Silica

Silica is not a normal constituent of the feed and feed ingredients. Normally it should not be present in feed ingredients. In fish meal it is permitted around 5% owing to its origin from the sea. But there is a tendency to adulterate the fishmeal with nicely grounded sand, just to increase the weight of the feed. Silica is a non nutrient substance and it dilutes the feed. Presence of silica in feed unnecessarily leads to waste of money. While procuring any feed ingredients it is advisable to check for sand and small stones and if present it should be rejected.

8. Antinutrients

Antinutrients are the factors which disturb normal digestibility of any particular nutrients. It is a common problem in the soyabean oil cake. In the raw soyabean oil cake, anti nutrients such as saponin, haemeagglutinin, trypsin inhibitor and tannin are present. However, these antinutrients are get destroyed when the soyabean are roasted. Sometimes inclusion of raw soyabean meal or ineffective roasting can affect the birds. So, while purchasing the soyabean oil meal the effectiveness of roasting must be checked.

9. Substitution

It is one of the ways of adulteration. Here, the feed ingredient of cheaper type of similar variety is mixed with higher cost feed ingredients. For example, the groundnut oil cake is nicely grounded and mixed with fishmeal. Fishmeal is of animal origin, and is having higher biological value because of the presence of unknown growth factor. Substitution of groundnut oil cake in the fish meal lowers the real nutritive value. Hence, it is advisable to buy the good quality dry fish and get it ground.

The poultry farmers are advised strictly to follow the above discussed facts while formulating mixing their own poultry rations or compound feed in order to maximise the profit margin.

❑❑❑

Chapter-44

Hazards of Ammonia in Poultry Farm

Micro environment of the house and in particular atmosphere purity needs to be given greater attention to keep up threshold level of management. Bad litter management along with poor cross ventilation increases the accumulation of the ammonia fumes level in the atmosphere. It is easy to assess the farm management standard by smelling the ammonia level in the atmosphere while entering the farm sheds. The causes for excess ammonia levels in atmosphere, hazards due it and suitable control measures are presented.

Causes

Ammonia gas is liberated from the litter due to microbial degradation of the uric acid present in the droppings. If the uric acid present in the droppings gets dried up in litter, the problem of the ammonia fumes is less. Problems of the ammonia in the farm increases only when the moisture content in the litter is high, thereby prevents drying of the droppings. Moisture content of the litter depends upon numerous factors such as housing, space density, ventilation rate, quantity and type of litter used, water content in the droppings, quality of the feed and water, relative humidity of the atmosphere etc. As the moisture content of the litter increases beyond 20% the microbial content of the litter is also high due to the favourable conditions. Mostly the problems of the ammonia are seen during rainy season and in winter season. Most of the farmers having layer farms close down the sides of the poultry houses with gunny bags during winter season to avoid chillness but leads to accumulation of atmospheric ammonia. This is unwarranted as the normal atmospheric temperature needed for the adult birds is 75°F and hence chillness of winter season is tolerable to the adult bird.

Hazards of Ammonia Pollution

It depends upon the ammonia level in the atmosphere and duration of its presence. As and when ammonia gas increases beyond 25 ppm in the microenvironment, it creates localized inflammation in the eyes and respiratory tract. Constant irritation of the eyes results in the lacrimation, conjunctivitis, sinusitis, coryza and the ciliary epithelium present in the respiratory tract gets damaged and favours entry of pathogenic micro organisms. Excess ammonia gas present in the atmosphere creates stress. Birds feel uncomfortable and show signs of respiratory distress. It leads to drop in production performance. As the condition worsens, it leads to maggots, nuisance in the litter, fly problems and if further deteriorates, it causes foul smelling. In the deep litter, the birds are having a dirty look. If the duration of problem of ammonia increases, birds become easily susceptible to all the viral, bacterial, parasitic and fungal diseases. Secondarily, it also produces breast blisters and leg boots. Overall the farm sanitation is getting worse. Lastly, it also irritates the farm labourers who losses interest to keep up farm management. It also causes nuisance to the people residing nearby and reports against poultry farm.

Control Measures

It must aim to spot the problems. It could be due to one of the many factors, which are as follows.

1. *Litter management*: It is advisable to use good quality litter material in sufficient quantity. The nature of microbes, temperature, water content and pH of the litter influences the fermentation rate and thereby ammonia production in the poultry house. Every attempt must be made to avoid the leaky water, spillage of water, leaky roof, seepage of water from floor etc. On any account, the

moisture content of the litter should not go beyond 20%. Easiest way to access the litter moisture content is, by taking the handful of litter and slightly squeeze, in this way, the good litter should not be caked or too powdery. In any place, there should not be any caked up litter. It is better to change the location of waterer frequently. It is advisable to change the litter with fresh ones, if it gets worsens.

2. *Ventilation flow*: The farm should have good cross ventilation facility. Normally, the recommended air velocity should prevail in and around the farm. Farm side wall should not be more than 18". There should not be any blockage for air flow. Cob webs and dust in the chicken mesh and welded mesh should be cleaned frequently.
3. *House Design*: The farm house should be constructed with scientific line. Orientation, length and breadth of the farm, height of the side wall overhang length, distance between the house, stocking density and distribution of equipment must be designed on scientific basis.
4. *Quality of feed* : The type, quality, nutrient density, energy level, mineral content, salt content, presence of laxative factors such as molasses, excessive potassium, low levels of toxins, low levels of microbes are accounted for water content of the droppings, and in turn on water content of the litter. If there is water droppings, the feed should be checked first for any problems.
5. *Water hygiene*: Quality and nature of the water is also other important factors which decides water content of droppings. Excess mineral content and salt content lead to watery dropping. Contaminated water and unchlorinated water can create mild low grade enteritis. Waterer either in cages or in deep litter should be kept in perfect hygienic condition. During evening time while watering, it is better to discard the contaminated water and filled with fresh water so as to lessen the chances of intake of impure water.

Even though there are remedial measures for lowering the moisture level in the litter and to decrease the microflora of the litter, it is always advisable to change the litter with fresh one, once it is caked up with releasing of ammonia gas. However, it is to be stressed the exact problem needs to be pin pointed and that must be solved.

□□□

Chapter-45

Hazards of Excess Moisture in Poultry Farm

Water is an important nutrient for all livestock. Presence of excess moisture in unwanted place creates numerous problems in poultry industry. Excess moisture in litter, feed and wind induces numerous health problems.

Birds require water for drinking purpose and by metabolic process. Excess unwanted water from bird's body is eliminated through egg, respiratory air and faecal matter. Presence of excess moisture in poultry farm causes many problems.

Excess Moisture in Environment

Presence of excess moisture in air causes so many problems. Presence of moisture in air acts as a nucleus for transport of micro-organisms from one place to another. Moisture is vital factor for reproduction of microorganism. Micro environment factors are equally important in the farm. Higher environmental temperature with high relative humidity makes the birds easily susceptible to the heat stroke. Where there is increase in body temperature in birds, the excess heat is eliminated by evaporation of water from lungs, so as to stabilise the body temperature. During the conditions of excess heat, relative humidity and environmental temperature, the excess heat cannot be eliminated through water vapours from lungs. Such circumstances make the birds easily succumbed with heat stroke condition. Micro environmental temperature of 75°F along with less relative humidity is more conducive for the birds. In the places of higher environmental temperature of 85°F, drop in egg weight, size, shell thickness, egg production are recorded.

In the winter season, higher moisture content and lower environmental temperature are seen. Presence of moisture in any place is conducive for luxury growth of all types of microorganisms. Multiplication of microbial organisms create disease outbreaks in the poultry farms. Chronic respiratory disease, infectious coryza and infectious bronchitis are such type of diseases usually seen in this type of excess moisture condition.

Excess Moisture in Litter

Farm hygienic conditions are easily assessed by seeing the nature of litter condition. Moisture content should not exceed 20% in the litter. Excess moisture in the litter leads to multiplication of microbial organisms and caking of the litter material. Excess moisture with fermentation of litter keeps the litter in warmth conditions, which are very conducive for growth of microbial organisms. Caking of litter inturn produces ammonia gas. Presence of ammonia in the poultry farm environment irritates the bird's respiratory tract mainly. Continuous exposure and production of ammonia mainly destroys the ciliary epithelium present in the respiratory tract. Ciliary epithelium is responsible for prevention of entry of microbial organisms in the respiratory tract. Destruction of ciliary epithelium makes the microbial organisms easily to enter and causes the diseases. Ammonia fumes should not exceed the level of 20 ppm. Level beyond this, affects production, health and disturbs the working tendency of the farm labourers.

Excess moisture in broiler farm, dirty the broiler birds due to continuous exposure of semisolid litter to the feathers. Longtime exposure of this conditions leads to breast blisters in the broiler birds. Breast blisters affects carcass quality. Excess moisture also favours the spreading of intestinal worms. Incidence of *Ascardia galli* and *Heterakis gallinarum* are increasing. Apart from this, domestic fly and other ectoparasitic incidence are seen. Overall, there is unhygienic atmosphere in the farm. Particularly in the caged birds, increased domestic fly nuisance are quite evident during excess moisture condition. Domestic fly spread diseases in the birds.

Excess Moisture in Feed

Production in the farm depends on the feed quality. It is advisable to purchase good quality feed. There should not be excess moisture in the feed. Moisture content in feed should not exceed 10% in summer season and 12% in winter season. Increase in moisture content in feed correspondingly increases the feed weight. It also increases unnecessarily the cost of feed. Presence of moisture in feed is conducive for growth of microorganisms and subsequent production of mycotoxins. Mycotoxins like aflatoxin B, citrinin, ochratoxin A are now become common. For luxury growth of fungus and resultant production of mycotoxin, requires environmental temperature of 21°C, moisture content in the feed of 13 to 15% and relative humidity of 85%. Mycotoxins are commonly seen in maize, groundnut oil cake and fish meal. Incidence of mycotoxicosis also hampers the antibody production and cause immunity breakdown. Owing to which, such type of affected birds are easily prone to other diseases. Birds are easily attacked by all type of diseases irrespective of fact that the birds are vaccinated or not.

It is critical to avoid presence of moisture in the farm premises, to keep clean environment and also optimum production. Some of the factors necessary to avoid problem due to excess moisture are.

1. Efforts should be taken to reduce excess moisture and temperature in the environment.
2. During the period of excess relative humidity and low environment temperature necessary alternative arrangements should be made to keep the required conditions.
3. Moisture content in the environment should not exceed 20% in litter.
4. Ventilation flow should be optimum.
5. Caking of litter and ammonia production should be avoided.
6. Moisture content in feed should not exceed 12%.
7. Spillage of water in farm should be avoided.

8. Farm house should be constructed as per design.
9. Disease outbreak should be controlled.
10. Dead birds should be subjected to post mortem and suitable treatment and preventive measures should be attended immediately.

□□□

Chapter-46

Litter Management

Management of deep litter finds an important place in poultry farming. Quality chicks, feed and scientific management are three important factors for successful poultry farming activity. In the poultry farming, after bird's sale, disposal of used old litter is critical. In the layer farming, replacement of litter is necessary once in six months.

After the removal of litter, the adherent litter material should be scraped and then subjected to thorough floor cleaning. Then the floor must be exposed to fire of blow lamps. Due to higher temperature, pathogenic organisms can be killed. Bleaching powder is having good germicidal effect hence can be very well used for cleaning. With an interval of one day, white washing of the walls and floor is quite necessary.

Litter material should be less priced and easily available at all times. Litter should easily absorb water and in dried condition. Litter should not be a place for production of living pathogenic organisms. Litter material should neither wet nor too powdery. Poor management of litter leads to outbreak of diseases.

Density of Litter

Depending upon the age of the birds in the layer farm, the height of the litter to be maintained in the layer farm varies. The recommended height of litter is 2 inches, 3 inches and 4 inches for to 8, 9 to 20 and above 20 weeks birds respectively.

Density of the litter should be altered according to the environmental conditions. Liberation of heat is a normal phenomenon in the litter. Increased consumption of water leads to higher moisture content in litter. Increase in litter material in winter season is advantageous. At any time, the moisture content of the litter should not exceed 20% level. Increase water content in droppings during summer session and lesser evaporation of moisture in litter during winter session, results in higher moisture content in the litter. In the winter session, raking of litter and removal of caked litter is quite important. If it is ignored, it is a place for multiplication of infectious organisms leading to disease out break and foul smell in the environment. In addition highly injurious ammonia gas accumulates and creates environmental discomfort to the birds. Ammonia production leads to respiratory disease in birds. Increase in potassium content and salt content in faecal material leads to increase in moisture content in the litter material and consequent caking of litter material. Quality of litter determines the efficient management condition of the farm.

Problems Due to Poor Litter Management

1. Litter management is critical even from brooder chicks. "Increase in water contents in litter in brooding period leads to adherent of litter material in the legs, toes and nails. Such condition is called booty legs".

2. Outbreak of coccidiosis in 4 to 12 weeks aged birds is mostly due to bad litter condition. Moisture content of 20% in litter materials makes the oocysts of coccidia, to get easily sporulated. Sporulated oocysts readily infect the birds. Mostly these infectious organism mixes with water or feed and induces outbreak of coccidiosis.

3. Presence of ammonia in the environment to the level of above 25 PPM irritates the respiratory system of birds and human beings; thereby the birds are easily prone to the respiratory infections.

4. Presence of blister in the breast region of birds is due to bad litter management. It also causes dirty feathers, which give trouble in the sale of live birds.

Hints for Litter Management

1. Adequate ventilation is necessary for good litter management. Cross ventilation in the poultry farm should be effective. Ventilation blocks should not be present any where.
2. Cobwebs and other dust materials should not be present in the mesh and walls.
3. At the time of increase in water content, the moisture content of the litter material should also be checked. Presence of molasses, potassium and salt in the feed are predisposing factors for higher water content in the faecal matter and subsequent caking of litter.
4. Wastage or spillage of water should be avoided during the placement of waterers.
5. There should not be any leakage in the waterers.
6. Caking of litter frequently seen in the place of waterers, hence periodical checking and removal of caked litter are quite essential.
7. Raking of the litter material before placing the waterers is quite appropriate. In this way, upside down of the litter material is feasible.
8. Raking of litter should be done slowly so as to avoid dust in the environment.
9. Caked litter should be removed then and there. Packing of the handful of letter in the hand can give an idea of litter condition. In this packing, litter should not be too powdery and not caked.
10. Addition of 8 to 10 kg of dried lime powder per 100 square feet can control the moisture content in the litter.

In these ways, it is necessary to maintain litter condition for better production and disease control.

❑❑❑

Chapter-47

Poultry Management in Summer

Poultry production in our country is expanding in to an industrial scale of operation since a few decades. Recent advances in poultry breeding leads to the development of high producing strains which are having commercial importance of exploiting hybrid vigour. Those birds of high producing nature are more susceptible to various stress problems, which lowers the profit margin due to higher percentage of morbidity and mortality. These problems are aggrevated during extreme hot season. Several problems are arising due to severe summer. They are,

- Managemental problems
- Nutritional problems
- Disease problems

Managemental Problems

The cage reared birds are more proned to stress during summer than birds reared in floors, since the space allowance provided is limited. Hence summer alleviative measures are to be taken for the cage reared birds. In cage houses effective cross ventilation should be provided with an extra care. Any obstacles preventing the ventilation should be removed from the sides of the cage houses. The water provided for the birds in extreme temperature period is heated up along with the cage and metal waterers. Hence cool, fresh drinking water with antistress vitamins should be provided.

Litter Depth

The litter depth or thickness of the litter should be maintaned at recommended level. It should be raked well periodically. If the litter depth is not kept as recommended the heat built up in the litter will be comparatively increased that adds the harmfulness to the birds in addition to the atmospheric heat increase.

Stocking Density

The floor space should be increased during summer season since the heat produced from the birds is more when the usual stocking density is followed. This heat causes more stress in addition to the environmental heat and hence it is advisable to reduce the stocking density by increasing the floor space allowance.

Ventilation

During extreme summer season the environmental heat causes a severe threat to the birds, in addition to the heat disseminated from the birds. The total heat built up in the poultry house should be eliminated quickly, otherwise the symptoms of heat stroke will occur in birds and hence it is advised to eliminate the total heat by installing pedestal fans / exhaust fans / providing additional ridge ventilation.

To reduce the heat radiated from the roof, the top of the roof should be painted with white colour and the bottoms of the roof should be painted with black colour. If the roof is of asbestos or tiles, it can be insulted by spreading paddy straw or thatches on the top of the roof. If the temperature is too high, sprinkling of water on the top of the roof should be advised. Wet gunny bags should be screened on the sides of the poultry houses. This will reduce the heat considerably.

Before constructing the poultry house, short growing plants and trees like soobabul, Agathi, neem trees etc, should be planted around the sides of the houses.

Nutritional Problems

During summer months, the birds consume less quantity of feed than the usual consumption. This will lead to less consumption of nutrients intake and thereby resulting in poor weight gain, drop in egg production and egg size. Hence it is suggested to increase the density of the nutrients in the ration. The energy level should be reduced and protein, calcium, phosphorus and all other important vitamin levels in the ration should be increased in extreme hot seasons. During the day time, the birds will consume only less quantity of feed, because of the hotness. Hence it is advisable to encourage the birds for consuming their maximum quantity of feed during the cooler parts of the day. Wet mash feeding is also advisable for the birds during extreme hot summer. Increasing the frequency of feeding and raking the feed often will increase the feed consumption during summer. Since the water consumption by the birds will be high during summer season. Provision of cool, clean, fresh and wholesome water is advocated. Ice cubes should be added to the water at extreme hot temperature period. During extreme rise of temperature incorporation of antistress vitamins like vitamin C in water should be advised. More number of waterers should be provided than the usual allowance. Frequency of watering the birds should be increased.

Disease Problems

General disease resistance of the birds is declining during summer season. Extreme temperature during hot season is the predisposing factors for the outbreak of the various diseases like Ranikhet, Infectious Bronchitis, Infectious Laryngeotrachitis, Infectious bursal diseases, Fowl pox etc, in poultry. Hence necessary precautions are to be carried out to check these diseases by preventive vaccinations, following strict sanitations and disinfection procedures along with proper medications.

In this article, several problems in relation to management, nutrition and diseases during summer are discussed. Based on this, the poultry farmers are advised to strictly follow the basic principles for improving their flock performance during summer.

□□□

Chapter-48

Poultry Management in Winter

Micro environmental factors like temperature, relative humidity and wind velocity are need to be favourable for conducive growth of the poultry farming. Necessary modification needs to be made in this regard. In addition, decrease in environmental temperature during winter there is a simultaneous increase in moisture content in litter environment and feed. It causes numerous problems. It is necessary to adopt necessary winter practices to safeguard the birds turn all odds.

Generally in India, during the months of October, November and December requires winter management practices. Microenvironment temperature needs to be maintained at required levels. Brooding of chicks require different environmental temperature depending on the age of the birds. Chicks require 95°F. During second week, chicks are comfortable at 90°F. In the third week, chicken demands 85°F and at the fourth week it is 80°F. Fifth week chicks fall well at 75°F. During the winter and rainy season, the temperature drops to less than 75°F especially during the night time. Variation in the environment varies with different places.

Body condition and physiological activity of the bird changes according to environmental temperature. Drop in environmental temperature affects the birds with wet condition. In this condition, the chicks huddle together under the electric bulb as to keep them in warm condition. Severe huddling leads to mortality in huge number. Inside farm house, temperature is very critical in winter season. Both during day and night times artificial warmth is compulsory by electric brooding. If electric bulbs are used for warmth, number and intensity of bulbs needs to be increased. Cold winds should not enter the farm house. Gunny bags should be hanged in proper direction to prevent the entry of cold wind.

Lesser age groups of chicks require preventive screen in all our sides, to keep up higher environmental temperature. However in adult birds, preventive screening must be in one side only, where there is much possibility for the entry of wind with velocity. This basic factor is important to keep the ammonia level in the farm within permissible level. Complete closure of both sides of farm house with screen like gunny bags in grown up birds affects cross ventilation. Effects on cross ventilation leads to accumulation of ammonia in litter.

Heaking and splashing of rain in the farm house must be totally avoided. Any problem in this regard have to be set right immediately.

Water Hygiene

In winter season, there may be simultaneous rains. In the rainy season, generally the drinking water is contaminated. All preventive measures have to be taken up to store drinking water properly. At any account, any spoilage of rain and waste water should not get into drinking water. Use of right disinfectant is essential. Easily available bleaching powder of 4 to 7 gm/1000 litres of drinking water with a contact period of half an hour is advisable.

Remaining water in the waterer is need to be discarded properly. Stagnation of water at any point outside the farm house should not be seen.

Litter Management

Litter is a critical factor in winter management. Moisture content in the litter should not exceed 20% at any time. Owing to the higher relative humidity in the winter season excess moisture present in the litter cannot get evaporated. This leads to caking of the litter.

In the waterer, the spilling of water should be avoided. Raking of litter is to be done daily. Such measure avoids foul smelling and ammonia irritation from the litter. Litter is a main place for multiplication of all infectious organisms. Increase in moisture content favours the multiplication of infectious organism. Ammonia production and consequent irritation of respiratory tract mainly leads to the out break of disease. There is drop in egg production.

Caged Birds

Accumulation of droppings under the cages are need to be avoided, such accumulation increase the domestic fly nuisance. Appropriate disinfectant should be sprayed over the litter. Increase in ventilation rate can reduce the moisture level.

Feed

Feed is an another important part which requires careful attention. Generally moisture content should not exceed 10% in the summer season and 12% in the winter season. Increase in moisture content favours the growth of fungus leading to production of mycotoxin. During raining and winter season environmental temperature is lower. On this occasion, it is not possible to dry the feed; mycotoxicosis causes many damages in the birds. As such, it is necessary to keep good quality feed.

Disease Outbreak

In the winter, the birds are exposed to stress due to low environmental temperature.Coccidiosis, respiratory disease and mycotoxicosis are such common disease in the winter season. It is imperative to take all preventive measures during the winter season to avoid disease outbreak.

Moisture content in litter if it exceeds 20% makes the oocytes of coccidia become infective by sporulation. Preventive coccidial medication in feed and water is advisable. Ammonia problem should be avoided. Antibacterial or antibiotics medication is necessary as a preventive measure of respiratory disease. Planned winter management practices protect the birds and keep up the production level continuously.

□□□

Chapter-49

Noise Management in Poultry

Poultry house environment comprises of temperature, humidity, air velocity, light and sound. The former four factors are controllable within the preview of the designer of the poultry house whereas the fifth factor is extraneous and may not be entirely controllable.

Sound is a periodic vibration in the air resulting in small pressure waves. When the pressure waves reach the ears, the vibration is picked up by the ear drums and transformed into perception by the auditory organs. If the sound is unwanted, it is called noise. Avian ear structure is similar to those of lizards and alligators except with the presence of auricular feathers. These feathers do not impede the penetration of sound waves.

Chicks can hear the sound frequency of above 400 Hz (Hertz) whereas hen can hear above 300 Hz. The audible sound for the human ear ranges in the frequency from 20-20,000Hz. So what is noise to man may not be noisy to the birds.

Extraneous sounds are quarry blasting, sonic booms, train noise, cracker noise, neighborhood industrial noise etc. There is some concern among the poultry breeders about the possible consequences of these sorts of sounds on the egg production. Apart from the extraneous noise, there is also panic and loss of production among birds due to vibration which is a low sound and can only be felt.

Effect on Behavioral Response

The sight of a depressed or agitated chick may not elicit any response of the indigenous hen. But "calls of distress" even out of sight from its offspring immediately receive the hen's attention. This illustrates behavioural response of the hen to the sound.

Effect on Production Performance

It may be expected that the depression in egg production occurs because of disturbance due to (i) irregular, intense and discontinuous noise, (ii) continuous, regular and frequent noise and (iii) vibrations. To pin point the kind of response access is necessary to production records of the flock with the dates on which noise occurs.

Habituation

If the drop in the production is thought to be due to noise, the first task should be to determine whether the repetition of the disturbance of sound is possible so as to solve the problem. This effort is called "Habituating" the birds. The warning of the fear causing fall in egg production is recorded as simplest sort of learning in birds and this learning is called "Habituation".

Habituation occurs automatically with the original sound whenever the sound is regular, frequent and continuous. Eg. Traffic noise, train movements etc.

But the Habituation does not occur whenever the noise appears intensely, infrequently and irregularly viz. Quarry blasting noise, cracker sound, barking of dogs and loud speakers nuisance etc. These sudden noises make the birds to frighten with resultant injury and drop in production. For these types of intermittent sounds attempt may be made artificially to habituate the birds by tape recording of these noises and playing back these sounds regularly to the birds so as to increase the frequency and regularity of the noise. In the remedial measures in correcting the production depress the following suggestions may also be taken into account.

1. Habituation does not affect the production when it occurs in grower stage. It may therefore be wise to acquire pullets which are exposed to noise i.e. reared in the same area in which they are to be housed.

2. Birds are extremely sensitive to vibrations, the receptors for which are found in hock joints. So, in diagnosing the problem of drop in production, the vibration may also be taken into account by feeling the vibration of the poultry house equipment.

3. When circulating fans are to be provided to reduce the temperature inside the pens, low RPM fans with sleeve bearings may be fitted. This produces less noise.

4. If the noise is produced by other units of the poultry house owner, the noise apartment may be planned by enclosing the source of the noise and preventing the radiation to the environment.

Summary

It is advisable to locate and design poultry houses away from areas where frightening due to noise is likely to occur. In case, such designing is not possible, habituation is necessary. Regular/frequent and continuous noise, habituation may occur naturally. To infrequent and irregular noises habituation may have to be done, by recording and playing back these noises at regular and frequent intervals. Habituating the birds may preferably be done when the birds are in grower stage. To the extent possible production of noise due to circulating fans can be controlled by installing low RPM fans.

However, further intensive research is required in this field of poultry management. Study can be made in designing the poultry house, where reflection, absorption and insulation of sound waves may be possible. It is also to be studied whether the disturbances due to frightening also predispose the birds to certain poultry diseases.

□□□

50

Hazards of Ammonia in Poultry Farm

In the poultry farm, surrounding environment and ventilation rate are quite important in maintenance of comforts of the birds. Bad litter management and poor ventilation predisposes birds for increase in ammonia content in the micro environment atmosphere. Concentration of ammonia determines the standards of the farms. Presence of ammonia causes many problems.

Causes

Microorganism present in the litter break down the uric acid in the fecal matter, leads to production of ammonia. Drying of the uric acid in the litter prevents formation of the ammonia. An increase in moisture content of the fecal matter is a basic predisposing factor for the production of ammonia gas, moisture content in the litter varies with farm house, birds density, ventilation rate, litter quantity, water content in the fecal matter and moisture content in the atmosphere. Increase in moisture content in the litter increases the microbial load in the litter; problem of ammonia is mostly seen during winter and rainy season. Closure of all the sides of the poultry farm with gunny bags during winter and raining season predisposes for ammonia production. This type act is unnecessary, as comfort temperature zone of the bird is 75 F.

Problem of Ammonia Gas

Intensity of the problem of ammonia depends upon its concentration in the atmosphere. Concentration of ammonia above 25 PPM affects respiratory tract and eyes. Continuous irritation causes lacrimation, eye irritation, blocking of nose and respiratory passage and destruction of ciliary epithelium, finally facilitating early entry of infectious organisms. Higher ammonia concentration also induces stress in the birds. Stress condition makes easily the birds fall for disease. Consequently there is drop in production. Condition worsens due to increase in domestic fly and maggots production. Fermentation of litter is accounts for foul smelling. In these conditions feathers of the birds get dirty and look ugly. Birds loose overall appearance. If the condition prevails for longer duration, there is possibility for outbreak of bacterial, viral and parasitic diseases. In the broilers, blisters in the breast region are commonly observed. Litter materials are also adherent to the legs. Irritation of ammonia gas also disturbs working conditions of the farm labourers. Increase in concentration of ammonia also causes inconvenience to the nearby resident.

Preventive Measures

Factors causing the ammonia production should be identified and solved.

a) Litter Management

Good quality and appropriate quantity of litter is desirable. Microorganism type, environmental temperature, level of moisture content, ionic concentration and chemical nature of litter are the factors governs the ammonia production. Leakage of water pipes, spillage of water, leakage in roofs, moisture in the floors, are to be located and arrested immediately. At any account, moisture content of litter should not exceed 20%. Quality of the litter can be determined by taking handful of the litter material. It is advisable to change the place of the waterer position, to avoid water retention in one place. Water should not be kept on one and same place. Good litter management is critical factor to control the ammonia level.

b) Ventilation

Good cross ventilation is necessary. Recommended ventilation flow should be implemented. There should not be any blockade in the ventilation passage. Weld mesh and cages should be cleared from cobwebs and dust.

c) House design

Scientific design of the farm house is necessary. Length, breadth, side wall height, distance between one house to another house, density of the birds, instrument place are to be arranged in scientific way.

d) Quality of feed

Feed type, quality, nutrients content, metabolizable energy, faecal quality, mycotoxin, microbial contents, lesser fiber content differentiate the moisture content of the litter material. Watery dropping should be checked, by seeing the quality of the feed.

e) Water quality

Water quality and content also determines the litter moisture content. Increase in mineral content and salt content causes watery dropping. Contaminated water is having a chance for inducing enteritis. Waterer in deep litter and cage system should be hygienic. Water in the waterer should be changed in the evening times. Water distribution needs to be checked for hygienic conditions. It is advisable to use soft water, to avoid salt sedimentation in the water pipes or erosion of pipe. Corroded pipes also contaminate the pipes. Water distribution points, water vessels, waterer should be cleaned very well.

Lowering of the moisture content in the litter correspondingly lowers the bacterial load in the litter. It is the easiest way to control microbial load in the litter and subsequent production of ammonia.

❑❑❑

ABOUT THE AUTHOR

Dr. D. Thyagarajan, Director, Centre for Animal Production Studies, Tamilnadu Veterinary and Animal Sciences University entered into his professional career as Assistant professor on 22.03.1980. He has completed 30 years of service in the different position as Assistant professor, Associate professor, Professor and now presently as Director, Centre for Animal Production Studies. He was awarded the prestigious national "Outstanding Extension Scientist" award for the triennium 1996-1998 by the Indian Council of Agricultural Research with citation for his dedicated work on open education network in animal husbandry enterprises in Tamilnadu. He was instrumental in getting nine external funded schemes to the tune of Rs.252.24 lakhs. He has released five useful technologies. He has standardized the package of practices for scientific turkey farming in Tamilnadu.

He has published 60 research papers, 95 popular articles, 10 books, 25 books edited, 4 booklets, 16 pamphlets and 6 e-lessons. He was the Principal investigator of 11 external funded schemes. He has adopted five villages under DBT scheme for integrated scientific turkey farming. He was responsible for massive open educational network programmes between 1990-1999 for 7893 rural farmers

He has guided three Ph.D., and seven M.V.Sc students and further guiding one M.V.Sc student. He has received best oral presentation prize to his scientific article in international workshop. He has received UGC book grant and Tamilnadu scientific book society textbook grant. He was instrumental for getting three prizes to Tamilnadu Veterinary and Animal Sciences University.

He has reported four specific diseases in poultry for the first time. He is well versed in the investigation of disease outbreak. Reported first time, "Incidence of *Salmonella weltevreden* 3, 10, n 26 in broilers" and "Hepatocellular and Cholangiocellular Carcinoma in ducks" for the first time.

He has recently organized the following two workshops

i) National Workshop on opportunities in Poultry Production Biotechnology.

ii) National Seminar on Emerging opportunities in Alternate Poultry Farming System.

He is a life time member in Indian Veterinary Council, World Poultry Association, Indian Society for Nuclear Technology in Animal Science and Indian Veterinary Journal.